FOREST

WALKING AMONG TREES

Matt Collins
Photography by Roo Lewis

PAVILION

First published in the United Kingdom in 2019 by
Pavilion
43 Great Ormond Street
London
WC1N 3HZ

ISBN 978-1-911595-26-7

A CIP catalogue record for this book is available from the
British Library.

10 9 8 7 6 5 4 3 2 1

Reproduction by Rival Colour Ltd, UK.
Printed and bound by 1010 Printing International Ltd,
China. This book has been printed on FSC® paper from the
Samson Paper mill in China.

www.pavilionbooks.com

Ash, Mulberry and the Wild Wood

THE HARDY ASH

Among the stately hybrid plane and lime trees of London's quiet St Pancras churchyard, a single ash rises above a circle of stacked graves. Its broad roots are partially visible, weaving between the grey stones, some of which have become swallowed beneath folds of fissured bark. When the headstones were originally placed in this curiously ornate fashion, back in the mid-19th century, they would have surrounded a mere sapling only a fraction in size of the tree that now towers above them. Over time, however, the ash bole has risen and swelled, clasping hold of the heavy tablets, causing them to lift, sink and split. In a litter of fallen leaves the two materials now coalesce – an enveloping of stone into wood, stitched by strands of advancing bramble. The result is a monument: a profound, semi-natural exhibit of life and death entwined.

The ash is known as the Hardy tree, named after the novelist Thomas Hardy on account of his involvement with the churchyard. The young writer had been working as a junior architect in London during the development of St Pancras Station in the 1860s. An extension of railway had been proposed that would cut through the ancient churchyard of St Pancras Old Church and, with Hardy's firm engaged in the project, the unpleasant task of reorganizing the burial site was delegated to him. Under Hardy's supervision graves were dug up, coffins uncovered and the interred relocated elsewhere; the tomb-less headstones that remained were subsequently gathered together and arranged in a circle with the young ash at their centre. Such a morbid experience no doubt contributed to Hardy's waning enthusiasm for employment in London, which came to an end shortly afterwards. Before the St Pancras ash had developed into a tree of notable size he had returned to the Dorset countryside of his youth and begun a new career. There, far from the fog and frenzy of booming industrial London, Hardy found the inspiration that would underpin his Wessex novels.

St Pancras Old Church lies tucked away behind the busy station of St Pancras International; a short walk from the British Library, where, on occasion, I have edited sections of this book. In moments of terminal distraction – when sunshine has pressed in against the high and narrow windows – I have found myself beneath Hardy's ash tree in the stillness of the graveyard. It is a place of pervading calm, rarely overcome with visitors, where birdsong mingles with rumbling train engines and the murmur of distant traffic. Now over 150 years old, the great ash matches its neighbouring limes in height and frame and has accumulated the decorations of arboreal maturity: mosses and lichen; stains of red algae and deep fissures that ridge the bark. Its command over the gravestones is resolute and powerful, phlegmatic even – like a statue of a victorious king with one foot suppressing a serpent. No source seems to confirm whether or not this shrine-like tribute to the natural world fulfils an intended design, though given Hardy's life-long literary affection for the countryside I like to think he'd have been pleased with the result. It is nature's apex flora casting green over the grey: a tree in defiant splendour enfolding humanity under its feet.

MULBERRY, ALDER AND THE EXPERIENCE OF TREES

Affection for trees came to me as an adult. I won't claim to possess an innate connection, nor allude to a childhood spent running wild beneath a leafy canopy. My suburban youth presented little in the way of regular contact with the kind of rugged, mysterious forests that might offer such a transcendental departure from the average boyhood. Growing up in West London I had the benefit of a great many parks and interesting gardens, and, as a result, encountered trees from all around the world, but there was no deep woodland on my doorstep, no opportunity to wander and become lost among tall trunks and sunken dells (then again, the virtues of 'getting lost' are something only my generation seems to have become obsessed with). When I was eight years old I could identify just a handful of trees: an oak for its acorns, as many children can; holly for the sharp foliage that was brought into the house each Christmas. There were trees I could not name but whose features I knew well: the leather-backed leaves of white poplar, for example, collecting in the corners of my primary school playground, or the sycamore's winged seeds that spun freely from the hand. Trees to me then were certainly beautiful and compelling things, at times even enchanting, but for the most part they remained individually indeterminable. There was one exception, however, the black mulberry that grew in our front garden, about which I could have talked for hours.

I would have described the mulberry's thick swollen trunk with its circular protuberances that made the tree easy to climb even with no lower branches for support. I'd have detailed its giant leaves, heart shaped and rough – rough like a cow's tongue or unpolished stone

– and edged all around with shallow teeth. The leaves spent each summer an ordinary green but by autumn became lucid yellow and then black, floundering on the floor like mushy, wet paper. I'd have praised the mulberry's branches, solid as roof rafters; you could walk right out along them without causing the slightest bend. Best of all was to be up there in mid- to late August, among the vivid and copious fruit turning from acid green to dark, raisin red. In September, when the berries were fully ripe and dropping, I would come down from the tree smeared in blood-like juice: it streaked my school uniform purple and left pink stains on my hands even after washing. What the pigeons didn't demolish would be picked and eaten, or thrown at a sibling as the ultimate provocation. Later in the year, rattled by winter winds, the tree's leafless extremities tapped at my bedroom window, and creaked softly from within the crown where two or more limbs leaned heavily on one another.

These memories all carry an enduring physicality. They are rooted in elementary experience rather than any theoretical understanding. Even now I can recall the feeling of that sharp, gnarled bark on my palms and fingertips; the sense of touch leaves an impression that far outlives learned fact. At eight, or indeed eighteen, I could not have told you from which part of the world the black mulberry originates, nor how it differs from a white or a red mulberry. I didn't know that our *Morus nigra* flourished in our London garden due to its sheltered and comparatively warm position. I didn't know either that the tree wasn't particularly old but merely *appeared* old, as is the habit hereditary in its species. But this is the nature of the human affection for trees; foremost it is tactile and rudimentary, however eloquently it may be expressed through poetry and descriptive prose. No intrinsic quality determines one type of tree more profoundly attractive than another, yet individual forms can be as colourful of character as a human companion; one does not fall for oak trees so much as for one particular oak. The black mulberry retains my affection for no other reason than it was the tree that grew in front of my childhood home – it may just as easily have been a maple or a walnut, as either could have produced a tree of equal individuality or have made as formative a climbing frame.

When I came to horticulture in my early twenties, I began with trees. Before I even approached the botanical nomenclature of flowers and vegetables I wanted to be on Linnaean terms with each of Britain's natives. This was simply 'starting with the big ones', as far as horticultural study was concerned, though there was something appealing about putting names to such familiar and commonplace faces. The cycle ride from my parent's subsequent home in Carmarthenshire to the National Botanic Garden of Wales, where I attended my first horticultural course, followed a quiet back route past damp tracts of mixed deciduous woodland. On mornings before class I'd hop the low stone wall with my copy of *Collins Complete British Trees* and set about deciphering the winter buds of ash, willow and hazel. It would be raining, almost inevitably, as this was south-west Wales, but as the identity of each tree gradually took shape in my mind so the landscape began to resonate with new interest. Knowing just a little about the trees in this familiar valley lifted a veil from the countryside that had hidden an entire dimension. The same views I'd known for years became altogether new; mottled sycamores stood out along the river by the house; blackthorns lit up the hedgerows; that blob of 'dead' forestry high up on the hillside was in fact a stand of deciduous larch.

Of all the trees acquainted with during this time however, it is the common alder, *Alnus glutinosa*, whose physicality wedged itself firmly into my memory. Beyond the purple buds, which grow like painted nails on outstretched fingers, or the bright orange heartwood at the centre of the trunk, it is the smooth, white-flecked young bark that stays with me, and the brittle spray of stems carrying male catkin and female cone on the same branch. While staying in Wales I planted numerous young alders, using them to build the bones of a garden in the mud-slurry surrounding my parents' recently rebuilt house. Seeing that alders grew profusely in this damp part of the country, I dug out saplings from a privately owned woodland nearby, stuck them in compost sacks and transplanted them along the new garden boundary. Whether this was a good move or not (my aesthetic appreciation of plants and garden design developed in line with my knowledge, which, at this stage, was still somewhat limited), the act of planting a tree, of negotiating roots into a hole, and then digging a bigger hole and trying again; of pruning away damaged stems and attaching a stake for support; these things go further than the page of any guidebook, or indeed the learning of a Latin name. I cannot pass an alder now without being reminded of wet Welsh clay.

When the romantic poet John Keats criticized his peer, John Clare, for the realism in his poetry, Clare's response was to suggest that Keats actually 'witnessed the things he described', implying he should get out there and meet the wildlife he professed to adore. In other words, the beauty of Nature is an assemblage of nuances revealed – often unassumingly – only at the source. This book, therefore, is about the experience of trees; of meeting them in the field, as it were, up close. It follows some of the routes through which other people have come to work with or alongside trees, or interact with particular species, and aims to explore something of the diverse associations that shape our relationship with trees today. Through first-hand encounters I hoped to expand my arboricultural experience, taking a variety of common trees and finding new ways to appreciate both their individual and forested contexts. This book is a journal of personal experiences, rather than a miscellany of facts, and has been laid out as a series of essays relating to ten particular trees. It is not an account of the extensive uses extractable from each of the chosen examples, nor does it attempt to draw generalized conclusions about their future applications. The wooded world is a vast and complicated subject and if you're looking for a thesis of arboricultural analysis, you will not find it here. I have written this book as half plantsman, half walker – which brings me to address the way in which its chapters have been constructed.

THE TREES

Ten types of tree make up the ten chapters of this book, labelled using common names predominantly familiar to the Western world; for example oak, cherry and poplar. Each of these chapters is then divided into two parts, exploring in the first an encounter, application or event relating to that particular tree, and in the second visiting a forest in which it grows. In many chapters the tree species may vary: for example, the chapter on oak looks at holm oak in part one and English oak in part two. However, the species in all chapters belong to the same genus, in this case *Quercus ilex* (holm oak) and *Quercus robur* (English oak) – both of these trees are known colloquially as 'oak' and both belong to the genus *Quercus*. The motivation behind this decision was to allow as varied a context for the trees as possible. A tree in a forest behaves differently to a tree growing out of a castle ruin, for example, or indeed its felled timber, which offers insights of an entirely different nature to that which is still growing.

There is no cohesive theme linking together the trees selected for this book other than their historical familiarity, if only by name, to a European and North American readership. At the outset, Roo Lewis and I compiled a list of roughly 30 trees and began looking for corresponding stories and locations that offered a range of interesting perspectives. We favoured the less obvious cases wherever possible: stunted oaks, industrial tree farms, desert juniper and wolves in UK forestry. The list was gradually edited down due to limitations such as travel constraints, seasonality and species diversification, and final cuts were made according to the resulting features. It was often the case that a tree would throw up an interesting story, but lacked a forest that was both compatible and logistically feasible – and vice versa. The final 20 features therefore range in form (broadleaf and coniferous, living and felled), habitat (desert, coastal, urban, mountain) and geographical location. Regarding the latter, however, all trees selected are found in the northern hemisphere and reside in a temperate climate, with the exception only of *Pinus canariensis* (in Chapter 1), which is found in the subtropics. Restricting the content to familiar trees is the simple reason behind this narrowed demographic; a eucalyptus of the tropics, though naturalized in places like Portugal and California, would sit somewhat uncomfortably among the others in the chapter list.

The act of physically visiting forests is central to the book's narrative. Each of the locations featured comprise differing component materials; as forests they *look* different, *smell* different and even *sound* different. In some cases they exhibit the bones of ancient woodlands. In the Niagara region of southern Ontario, for example, that appears in Chapter 2, the vibrant autumn display reflects the original Carolinian Forest. There are woods too that are recent by comparison, having either naturalized – such as the sweet chestnuts of the New Forest in Hampshire – or become invasive, like Oregon's western juniper, whose resilience to drought has allowed it to gradually infiltrate areas of desert prairie. Being able to walk through these forests ourselves has set the platform for this book, informing the ways in which Roo and I have sought to interpret their various and contrasting attributes.

Collections of trees, unlike single specimens, are of a duel construct: they are both physical and cerebral. While their appearance is dictated by a specific floral make-up, geographical positioning, elevation and historical use, in the mind forests enter the realm of unrestricted imagination. Very few of us can walk through a forest absorbing only its materiality, discarding the innate responses it prompts within. We experience excitement, awe, trepidation and even fear; emotions equally instinctive as they are irrational. Forests and woods have long inspired such responses, which stem back in the reactionary mind to an age when tree cover was far more prevalent, and the dangers therein more threatening. They are the setting for our happiest notions of freedom and most terrifying nightmares alike. Literature thrives on the forest: Shakespeare's *A Midsummer Night's Dream*, Roald Dahl's *The Minpins*, Maurice Sendak's *Where the Wild Things Are*. Indeed, *The Wind in the Willows* might not have gained such prodigious popularity were it not for the vividness of Kenneth Grahame's Wild Wood. When hapless Mole goes in search of the illusive Badger, the wood is described as lying before him, 'low and threatening, like a black reef in some still southern sea'. The dark corners and ill-lit dells inherent in woodlands have inspired centuries of curious stories and folklore: the fable of the Babes in the Wood, the Ents of Tolkien's Fangorn Forest and many a feverish tale concocted by the Brothers Grimm.

In the mind a forest needn't be restricted by size. I love the description of Tove Jansson's Magic Forest in *The Summer Book*; a stand of only a few spruce trees at one end of her tiny island off the coast of Finland. In the book this little haven becomes an enchanted mass of dead branches; the realm of animals never seen but heard with rustling wings and scurrying feet. This is the lure of trees; when collected together they form an environment often too enthralling and characterful to be lacking exhilarating fauna, whether real or imaginary. When I first read Lydia Peelle's short story, *Phantom Pain*, I fell under the spell of the creature she conjured stalking the woods outside a small American town. As beloved pets go missing and possible sightings whip up a frenzy, hysteria settles on the town's residents leading them to question, 'can those woods out there – on the outskirts of urban sprawl – really still support a big cat like a panther or cougar?' Though in recent history the world's forests have seen a rapid and substantial decline, there is still something in us that wants to believe in the Wild Wood of bygone days. William Burroughs put it well in 1986 when he reflected: 'The magical medium is being bulldozed away ... the medium in which Unicorns, Bigfoot, Green Deer exist growing always thinner like the rainforests and the creatures that live and breathe in them. As the forests fall to make way for motels and Hiltons, the magical universe is dying.'

PHOTOGRAPHY AND TEXT

Forest is a collaboration; a book Roo Lewis and I discussed often during trips at home and abroad while working on projects together. The idea was always that both text and photographs would tell combined yet independent stories, making room for our own interpretations of each subject. In publications where images have been brought in to match neatly with prewritten text a little of the magic is inevitably lost. I like the notion that collaboration instead leads to a dialogue, as described by John Berger when working with the photographer Jean Mohr on *A Fortunate Man*. Berger expressed their process as a kind of conversation, 'building on, rather than mirroring, one another'. Though I wouldn't dream of comparing our efforts to such prestigious work, this is a little of the sentiment behind our chosen approach. Over the course of a year Roo and I visited the trees and woods, most often together though sometimes separately, and would, at the end of each trip, disappear into our respective domains to process the resulting material. The outcome is that Roo is little referenced in the text, just as I rarely feature in the images, though in bringing our efforts together two perspectives colour the one.

JOHN STEWART COLLIS AND THOMAS HARDY

In contributing to the well-documented subject of trees and forests it would be impossible to sidestep the great wealth of poetry, essays, prose and fiction already serving the theme. While a number of these works are referenced, one particular text is mentioned a number of times in this book and therefore ought to be properly introduced. It is, in my opinion, an account of a forest unparalleled in quality and insight, despite being penned by a comparatively little-known author. In the early 1940s, with the rest of Britain and the wider world preoccupied by war, John Stewart Collis – writer and biographer – began work as a forester on a private estate in Dorset. His task was to clear a stretch of woodland belonging to the estate owner, freeing its ash trees from a tangle of honeysuckle and ivy and thinning out those not profitable as timber. This was not the first time the writer had exchanged academia for this kind of earthy, hands-on vocation; he had already worked in a variety of roles on farms in the West Country. Across long stints in the ash wood however, Collis got to grips with the art of forestry, learning as much about his enjoyment of this strenuous sylvan activity as of the trees themselves.

It is a great shame that Collis's writing did not gain wider acclaim, though some of it does remain in print, notably *The Worm Forgives the Plough*, which brings two of his books – *While Following the Plough* (1946) and *Down to Earth* (1947) – together under the one title. The latter was itself already a book of two halves, the first recounting his time as a farmhand and the second as a forester. The forester section is called *The Wood*, and it is this text to which I frequently refer. I would urge anyone interested in trees, woodland and forestry to read this book, and to enjoy, as I have, its many profound observations. Across a series of short essays Collis beautifully articulates the story of a working wood: its faunal inhabitants, its changing floral carpet and exhibitions of life and decay. *The Wood* is a blend of theoretical, philosophical and physical responses to trees and woodland, all of which result from the fundamental application of what gardeners, farmers and foresters would call 'a bit of hard graft'. 'For me it is first the tool, then the book', Collis proclaims, making him, in my opinion, an instantly likeable writer. At its heart is a narrative of wonderful simplicity; a man, with his tools, working in a wood.

Collis had only one biographer in his lifetime: journalist and friend, Richard Ingrams. In his memoir of Collis, Ingrams asserts that, 'no reader of *Down to Earth* could miss the note of exhilaration sounded during his time as a forester, during which he was able to alternate strenuous physical work and writing. He thought, as he said, that he had discovered the ideal existence.'[1] I have enjoyed writing about trees every bit as much as I have enjoyed working with them: none are the same, yet all communicate stories, wherever they happen to grow. In many ways Thomas Hardy's ash tree and Collis's ash wood establish the structure for this book: one part story, one part wood, both the same tree.

Pine

So familiar are we with pine trees that 'pine' often is a synonym for 'conifer'. With over 100 different species occurring in the northern hemisphere, it is the pin-up of the evergreens, with species designated as national trees in countries as geographically scattered as Afghanistan, North Korea and Scotland. Nine US states have pines as a state tree (Nevada in fact has two: the single-leaf, *Pinus monophylla*, and the Great Basin bristlecone, *P. longaeva*). There are the stout white pines of Japan (*P. parviflora*), the durable, erect red pines of eastern Canada (*P. resinosa*) and those occupying high altitudes in Russia, such as *P. sibirica* of the Ural Mountains. Rome has its iconic stone pine umbrellas (*P. pinea*), while the arid heights of the Rocky Mountains are characterized by their contorted limber and lodgepole pines.

Beauty is of course derived most prominently from the pine's distinguishing needles and cones. Species have between one and seven needles per cluster (fascicle), splayed outwards from buds along narrow shoots. The cones also show variation: some are elongated like those of the sugar pine (*P. lambertiana*) or large and heavy, like those of the big-cone pine of California (*P. coulteri*) – in the green, *P. coulteri* cones can weigh up to 11lb (5kg), resembling large, spiky pineapples. As a fast-growing timber, we employ pine for our floorboards, tables and chairs; a commercialized wood made all the more pleasant by the sweet aroma locked within.

Scots pine (*Pinus sylvestris*)

Return to Caledonia Sylva

There is a forest up there on the heather-red hillside, knee-high and infant, crouched beneath the tangled and wind-rushed scrub that blankets the Highland moor. It is a forest 6,000 trees strong; of birch, rowan, aspen and alder, with scattered sapling pine springing like bushy horsetails from clods of upturned soil. It is a forest of new beginnings bearing the pattern of an historical fabric; a fabric that, like dynastic tartan, binds identity with a distinct set of colours and shades. An initiated, planted forest it mirrors something grand that went before it; an arboreal pledge, made by one generation for the inheritance of another. From where I stand however, at the foot of the hillside in the curve of the River Moriston that drains Loch Cluanie eastwards into Loch Ness, the view remains a scene of customary moorland; all heather, bilberry, spongy moss and spent golden bunchgrass, a familiar mosaic of ochreous tones, blended one into the other. A sharp spring sunlight enlivens their thatched surface with heightened colour, yet the little forest remains hidden low within, sending new roots quietly down into the south-facing soil somewhere between me and a great white cap of montane snow above.

I've come to Dundreggan Estate in north-west Scotland; 10,000 acres (4,047ha) of wild sprawling moor and parcelled semi-ancient woodland, purchased in 2008 by the organization Trees for Life. Now one of Scotland's leading conservation charities, Trees for Life aims to reinstate a proportion of the nation's revered yet dwindling Caledonian Forest to the vast undulated terrain it is believed to have once commanded. The western Highland Glens constitute an iconic and arresting landscape – wide-open panoramas that have long stirred hearts and minds – but it is not the landscape that once was. Five thousand years of periodic deforestation, through cycles of fractious interplay between humans, animals and trees, have all but eradicated Scotland's former wooded wilderness and led to what Trees for Life founder Alan Watson Featherstone calls an 'outdoor museum', where the lignified stumps of former pinewoods lie exhibited as relics, preserved in the peat bogs that channel throughout the moors. The Great Wood, as the Caledonian Forest is also known, that once covered much of the central Highlands, is today restricted to isolated pockets of ageing old-growth pine and birch. These fragments are thought to represent just 1 per cent of the forest's historical distribution, raising concerns that, with many of the Scots pines themselves reaching the end of their natural lifespan and their seed-sprung successors at the mercy of unrestricted deer grazing, the Great Wood might slip away altogether under a sheet of advancing heather. Bolstered by the active support of volunteers, the charity has spent almost 30 years working towards Caledonia's gradual return, planting nursery-grown saplings into designated moorland areas and fencing out the herbivores that have previously decimated all signs of new growth. It is a project with a proclaimed '250-year vision' that looks to re-establish a self-sustaining and balanced ecosystem, where former inhabitants repatriate below a dappled canopy and seedlings may develop uneaten.

To reach the current planting area at Dundreggan one passes first through the 'aspen burn', a small acreage of unplanted, semi-ancient and relatively well-preserved pinewood. Occupying rugged ground rising onto the hillside slope, the wooded escarpment paints a fairly accurate picture of the Highland landscape that has been so radically reduced over the years. It is airy and light; a patchwork of lichen-swaddled birch and clusters of veteran pines that preside over protrusions of local Moine schist. It is moss-covered and stream-laden too; grey-green-black, a place where pine martens bound through arching bracken and treecreepers trawl long tails, like inquisitive mice, over deep fissures of pitted, vertical bark. The floor is uneven with scatters of brushwood birch, giving way here and there to great-mounded juniper and the bright buds of delicate primroses. But this enchanted margin is a mere vestige of a missing wood, the Great Wood, that in the national consciousness lies somewhere at the crossroads between legacy and legend – for the true nature of ancient Caledonia is a subject often debated, and like all ancient forests its roots are tipped with inherited romanticism. Some suggest it was a vast and continuous covering with hardly a glen not roofed with trees, while others, including the late woodland historian Oliver Rackham, argue a more fragmented network of woods broken up by burn and bog.

Caledonia, as legend has it, began with the Romans who in their upward expansion of Britannia were said to have been stopped short by a rough and impenetrable forest; a Boreal-esque jungle of bears, wolves, lynx and boar, and imposing pines harbouring fearsome and face-painted natives. The northern wilds of 'Caledonia sylva', as they named the region, gained a foreboding reputation in the imagination of the wider Roman Empire, carried all the way back to Rome itself as justification

for the relinquished imperial advance. These initial embellishments, however exaggerated, are thought to have shaped the lens through which the ancient forest is still viewed. As the contemporary Scottish nature writer Jim Crumley put it, 'the Great Wood of Caledon has the same fabled ring to it as the Loch Ness Monster'.[1] The rudimentary facts, however, most can agree on: that a woodland community of pine-birch-juniper developed gradually in Scotland from retreating ice 10,000 years ago, and what remains now is precious not only for its unique character but as an intrinsic and historical property of the wider Highland landscape.

Doug Gilbert, operations manager at Trees for Life, is showing me the route up towards the current planting site. Together we navigate a barely decipherable path through Dundreggan's mature wood, traversing a carpet of cowberry and blaeberry (the Scottish designation for bilberry, *Vaccinium myrtillus*), and leaping periodically across the weaving burn. Early this morning a team of eager volunteers drove into the hills at the high end of the estate, cut across the moor and began digging in new saplings. 'They're quite a walk from here,' Doug warns me, 'but you'll get a good picture of the differing terrain on your way up.' He'll take me along the aspen burn and out onto the moor, he says, at which point I can continue up through the glen until the volunteers – in their high-vis vests – materialize somewhere just beneath the snowline. Having spent most of his working life in conservational roles for organizations like the RSPB, Doug walks with a manner of mindful distraction, tuned acutely to subtleties within the surrounding environment. He stops at intervals to highlight some of the features characteristic of Caledonian woodland: rowans festooned with honeysuckle, dense layers of floor-sprung moss and – before I step over it – a cluster of busy wood ants warming in sunlight on a dry heather knoll. They cluster to absorb heat, he explains excitedly, transferring it down into the nest so as to charge the colony batteries at the outset of spring. Wood ants, like the resident songbirds, are a crucial predator in the broadleaves of Scotland's forests, preying on invertebrates (moths, caterpillars) that, unchecked, would defoliate the trees. Theirs is a key role in the balance of the wood. Our path soon enters a gathering of birches, a flurried mix of both silver and downy, varying in maturity and fungal ornamentation. Their leafless late-winter canopy is brittle and sparse, almost salmon-pink in colour. We exit to a contrasting view: a little way further, overlooking the steep river rill, stands one of Dundreggan's ageing pines.

Garden writer Hugh Johnson, in his great arboreal tome *Trees*, wrote that, 'few would deny pride of place to the Scots pine among the pines of Europe – even perhaps the pines of the world. The richness of its colouring, its wild poise set it apart'.[2] These are entirely indisputable observations. However, in the context of this particular forest, 'setting apart' is an unnecessary analytical measure;

in Caledonia the Scots pine stands all but alone. In the boreal forests of Eurasia, Scots pine mingles happily with tall spruces, firs and indeed other species of pine, but following the last Ice Age none of these cohabiting conifers made it onto British soil. We have only juniper (*Juniperus communis*), yew (*Taxus baccata*) and the Scots pine, whose natural range moved gradually northwards as the climate warmed, restricted now to Scotland and the eastern tip of England. And so to see here a solitary green giant occupying a relinquished space between copious birch, scattered aspen and gin-heather scrub, how could such a tree not leave an impression of magnificent supremacy? More so when you take in the details: the great dome of green-blue foliage; the fire-orange limbs in the upper storeys; the stout, stately bole of snakeskin bark. Little wonder it is the national tree of Scotland.

Doug puts the specimen at 250–300 years old; a similar vintage to many others on the estate. 'It's been a long hard winter,' he says as we make our way over to it, drawing attention to a number of fallen branches lying at the foot of the tree. 'They were weakened by the heavy snowfall this year and have since been shedding.' I clasp a spray of cold needles in a cold hand – they are still fleshy and aromatic, albeit limp and losing turgor, like woollen tassels at the end of a robust rope.

While Trees for Life approach reinstating the Caledonian Forest by planting nursery-propagated whips, this is only carried out in areas lacking mature trees and the reach of their wind-borne seed. The preferred method is to protect areas adjacent to existing woodland with deer fencing, allowing seedlings to prosper naturally in the absence of over-grazing. Intervention in this case is minimal, and the results far more organic. But excluding deer from existing woodland is only realistic in limited areas – Scotland in fact has the largest population of red deer in Europe, concentrated mostly in the western Highlands and Islands. 'They've had a place in this environment for a very long time – there's no doubt they want and need to be here. But with such a high density having been supported for so many years, and a lack of natural predators to control their numbers, we've struggled to get any tree regeneration.' Red deer will eat almost anything that is green, which, for the pine, means tips, needles and all. During the cold months of winter the woods offer a protective environment for Highland deer; a time when pine saplings are at their most conspicuous and vulnerable, exhibiting green while all else has deepened in colour.

Doug leads us out of the woods and a short way onto the open hillside. No longer intercepted by trees the temperature takes a palpable drop; a sharp breeze is up and battling at my ears, nose and bones. The path, if obscured before, now submits entirely to an enveloping of heather, bracken and juniper: the route ahead becomes a choice between slips of black bog and footholds made uncertain by matted foliage. Doug leaves me with a point in the right direction and an OS map of

the glen, should I find my way to getting lost. He heads back towards the wood, stooping to inspect the tips of young birch in the undergrowth. I rest a moment with the spectacular view. The hills opposite – those across the river facing north – reveal contours of contrasting topography: wavy strips of yellowed grass, spruce plantation, birch glade, more plantation and a final layer of snow. A handful of solitary Scots pines stands in the blanket of white like single oaks in a waterlogged field. Their crowns are broadly domed, just like an oak. As Doug says, from up here you get a sense of the mixture of topographical components that shape the Highland landscape, the full spectrum of past and present interactions with the land. A skylark emerges from the thicket beside me and lifts seamlessly onto the wind, carried steadily away like a fallen leaf on the current of a flowing stream.

By the time I reach the work party my nose is streaming steadily and my feet are sodden from sumps in the glen. I'm greeted by Group Leader Dom Andrews, who congratulates my apparently successful orienteering and introduces the volunteers. The site is encircled by an enormous and robust deer fence, inside which stand marker flags indicating the current planting zone. The ground between them is mole-hilled with clumps of earth upturned by a tractor ahead of the dig, and out of each protrudes a little whip of tree. The whips have been propagated at Dundreggan's onsite nursery and follow a ratio natural to the Caledonian Forest: the bulk are birch, which will get growing quickly and provide shelter for the establishing pine. On the ground, however, the planting is purposefully sporadic so as to avoid a formulaic pattern. There are long drifts of birch intermixed with dotted aspen; and here and there flashes of bright, juvenile pine show up against the dark earth. Planting directly into the grass, Dom explains, would mean a difficult start for the young trees and by turning the heavy clods they access instead an underlying soil that is naturally rich and peaty. Indeed, when later I'm handed a little downy birch of my own to plant, plunging a spade into the sticky earth feels immensely satisfying. But you have to admire the spirit of those dedicated enough to spend a succession of days up here in the wind, rain and, quite often, sleet. Volunteers come from all over, Dom says, sometimes from oversees, and spend a week planting at Dundreggan. If the weather forecast looks truly disruptive they might retreat to the nursery, but otherwise they'll be up on the wild moor. There is a tangible sense of purpose that pervades the team however: an obligation to put something back that future generations will live to enjoy. By 2012 Trees for Life had planted their millionth tree, now at 1,500,000 they're on track to double that number.

The next day, before taking the road back to Inverness for the journey home, I pay a quick visit to a small section of estate land planted by volunteers in 2010. The young grove of trees lies at the south end of Dundreggan, close to the river and accessible via a roadside gate. Eight years of deer fencing and careful monitoring have proved enormously successful here, and a forest that began as little sticks now rises to just above head-height. Having walked through an inch of enchanting remnant Caledonia it is no leap for the imagination to picture what this plot may one day become. One might picture too that in 250 years from now, the grove will have bled at the boundary and connected with others across the Highlands, like drops of seeping ink converging on a paper tissue. Referring once more to the words of Jim Crumley, '... the way ahead is slowly growing green again, and the Great Wood stirs from a long and ominous slumber and begins to throw new shadows in the resurrected sun.'[3]

Tamadaba Pine Forest

I n making sense of an island for the first time – building a picture in the mind of its geography, structure and identity – plants provide a reliable pathway. Exploring new ground I am always hopeful of recognizing a few, whether through previous encounters in natural landscapes and gardens, or from the pages of a guidebook leafed through on the journey over. Flowers, shrubs, trees become orientational anchors in a new environment; they are geographical waypoints but also characters representative of the locality, as weathered island natives themselves. Of course any unfamiliar landscape may be interpreted via its floral components, but there is something more accessibly comprehensible about islands – small islands – given their restricted mass and inflexible sea wall. Enclosure gives to plants a sense of unique and tangible prominence – it encourages deeper roots with which to clasp a finite soil.

Arriving on Gran Canaria, in the diminishing dusk fast falling to night, and winding a narrow, almost vertical road into rising hills, I find myself looking to the passing verges for glimmers of revealing vegetation. Plants appear in momentary flickers; huge sow thistles, *Euphorbia* domes, bright yellow globes of *Ferula* fennel and splayed pink mayflower; first impressions illuminated in the headlights. As the centre-most isle of the Canary's subtropical archipelago, Gran Canaria may be associated with packaged beach resorts and mass-market tourism, but these things occupy only the warm-watered periphery; the interior holds something quite different. I'm anxious to get up there – past the motorway, billboards and streetlights, and the banana plantations sheltering under grids of polythene sheeting. The road ascends quickly into a settling fog that, like a veil, blurs the transition from the garish into the serene; a few hundred yards upwards and even the dampened verges fade behind a screen of ocean mist. On the X-axis Gran Canaria possesses a diameter of a mere 31 miles (50km) and to circle it leisurely by car would occupy only a day. On the Y, however, it lifts to almost 6,500ft (2,000m) above the big blue Atlantic, summiting in the clouds at Pico de las Nieves and the smooth lava plateau of Roque Nublo. Historic volcanic activity on the island has inflicted great fissured vents and gorges that provide niches infiltrated by a diverse flora, and so to negotiate the interior is to traverse one ecoregion after another, each defined by contrasting plant life. To the west there are ravines of rock-sprung houseleeks and Paris daisies, crossing into mountainside scabious and broom; laurel woods in the north conceal endemic orange foxgloves and communities of leafless, cacti-like spurge – there is even a sand dune desert in the south, dotted with defiant, feather-leafed tamarisk. But towards the elevated centre, falling to one side of the uppermost hill town of Artenara – where I am headed – sprawls the Tamadaba pine forest for which I've come, comprised of *P. canariensis*, a champion of the Canary's endemic flora. Endemics, after all, are an island's ultimate natural charm, and Gran Canaria has more than its fair share.

Canary pine is the dominant tree in the western islands of the archipelago, forming forests that thrive across lofty, mountainous altitudes. As a once abundant native possessing durable, quality timber, the tree was always at risk of potential exploitation through commercial logging, and sadly on Gran Canaria, as on Tenerife, pine forests were significantly reduced in size before legislative protection was brought in to conserve them. By the 1950s it was observed that deforestation had destabilized an already volatile, permeable soil, leading to significant erosion and degradation of the hillside watershed. A subsequent governmental drive to replant the forests saw large areas reinstated – in fact, pine wood reforestation on the Canary Isles has been something of a remarkable success story, not only halting but going a long way in reversing years of arboreal mismanagement. The pines were quick to repatriate lost ground, in part due to a speedy growth rate, but also because they are so very much at home in these higher mountain terrains, even in areas of consistently low annual rainfall. Their tolerance of the dry ridges is supported by the rolling ocean mist that condenses on the pine's three-pronged, elongated needles and drips down to the roots, triggering thousands of successive seedlings to germinate. This is why Tamadaba feels so palpably saturated when, the next morning, I make my way in among its trees. Moisture literally hangs in the air here, trapped by the stillness of the forest. It sheens on bark surfaces and sparkles from drooping beard lichens, cooling and purifying the air. Your lungs draw deep for it. The mist is changeable too and in constant flux; at times vision retracts to only a few trunks deep, framing those closest against a colour-drained backdrop, like a film set almost, or from my own experience, the coastal woods of Oregon, where Pacific mist can in a moment swallow entire hemlocks and towering spruce.

In this environment, where little grows well beneath the evergreen canopy and carpet of slowly decomposing needles, whatever bears colour contrasts conspicuously with the gloom. The same endemic sow thistles (*Sonchus acaulis*) I met on the night's journey up, with their big, velvety dandelion heads, draw the eye between trees. As do the enchanting pink spires of *Asphodelus ramosus*, which, being springtime, form profuse drifts over the forest floor. Deeper into the forest there are more surprises, particularly where the terrain gets steeper. The silvery-blue, melianthus-like leaves of what appear to be Polido's burnet (*Dendriopoterium pulidoi*), a rare Gran Canaria endemic, spread out over the basalt rock. The archipelago's own subspecies of great spotted woodpecker drums out invertebrates in a pine overhead, and a chaffinch – perhaps the infamous blue-coloured native – calls repeatedly nearby.

It is something of a relief, after the closeness of Tamadaba, to get back up onto higher ground, following the pines to the island's central peak of Roque Nublo. Up here tree cover is less dense, broken by outcrops of rock sculpted by a volcanic eruption four and half million years ago. Nublo itself – the 'Cloud Rock' – sits at the top, a 220-ft (67m) monolith of freestanding stone. Caves formed beneath the flattened summit offer views out over a sea of re-establishing forest, and at the entrance of one I find an ideal seat from which to enjoy the evening sunlight. Its steady heat is a reminder that less than a hundred miles of ocean separate this island from the African mainland of Western Sahara. And this is where the garden plants are: the subalpines of the sun-baked rocks; the echiums, erysimums and argyranthemums British gardeners so determinedly cultivate for Mediterranean summer colour. The three intermix beautifully in natural communities down Nublo's hillside, catching light in outstretched blooms scattered in glades before the trees. To sit in sunlight at the edge of a forest, with its entirety in view, brings to mind the writer-woodsman John Stewart Collis and his daily breaks from employment in the Dorset ash wood. An appreciation of the sun verging almost on the religious necessitated what he deemed 'precious moments' out from under the canopy. They were moments to free the mind from bitter thoughts of the outside world, he declared, 'which often pursued me into the wood like loathsome hounds.'[4] Loathsome hounds one might picture in the foggy depths of Tamadaba, but up here the sun overrules. From this peaceful, elevated ledge I watch a billow of pollen blow free from the male cones of a periphery Canary pine, set loose by the heat. The sulphur-coloured cloud moves as a single entity, drifting lightly away to the east.

Hornbeam

There are over 30 species of *Carpinus*, which belongs to the same family as birches, hazels and alders. Hornbeam species are almost exclusively found in the temperate regions of the northern hemisphere. The European and American species are perhaps most widely appreciated for their spectacular forms and their suitability for municipal planting schemes, both as standard specimens and rugged hedging. The name 'hornbeam' is derived from the density and sturdiness of the wood, like horn. Their trunks and limbs tend to be smooth and twisted. Hornbeam leaves are sometimes mistaken for those of beech, as their overall shape is similar. However, the two may be distinguished by their leaf margin: beech leaves have smooth rather than finely toothed margins.

Hornbeam (*Carpinus betulus*)

The wood is a clamour of chainsaws, a racket that rises steadily to a crescendo before coming to a sudden stop. Sean has been grinding his way through the contorted limb of a hornbeam tree, the teeth of his saw duelling furiously with the dense heartwood at its centre, and in an instant the bow surrenders – moving barely an inch – and the saw has become trapped mid-cut. The chain bar is pinched tightly by the branch's rapid shift in weight, and, after an assertive yet conclusive tug, Sean steps back to assess its recovery. Richard, silencing his own chainsaw and laying it on the ground, walks over to join Sean in the solving of this predicament. The pair exchange a customary smile and together circle the tree. A trapped chainsaw is part and parcel of forestry work and is by no means a freak occurrence: living trees contain living wood, the will of which can be difficult to predict at the best of times. However, this particular hornbeam limb, which stems from a large stool that has seen many years of repeated coppicing, is fused with another, having wrapped itself like rope around a limb of equal girth. There were nine such branches to this tree when Sean began cutting it, most of them around 6in (15cm) in diameter and 30ft (9m) tall, and he has consciously left these two until last, knowing what difficulties their intertwining might present during felling.

Standing a little distance away from the action I watch the woodsmen perform a familiar procedure. The small team of volunteers, whose work party I've joined for the morning, also turn from their tasks to watch. Having tried to bend the fused limbs backwards to no avail, Richard walks over to a pile of stacked logs and returns with a length of sawn birch. He launches it at the upper stems, just above the chainsaw, upon which it thuds sending a small ripple up through the tough, rigid trees. Their tips in the understorey canopy sway a little, but the force of the knock is absorbed almost entirely by their heavy wood; hornbeam is so extremely tough that such a gesture barely marks its bark, let alone rocks the trunk. Richard tries for a second time to shift the upper limb, and then a third and a fourth. At the last strike one of the two limbs is abruptly freed from the stool, shifting very little yet transferring its entire weight onto the one with the trapped saw. A final hurl of the lumber by Richard and the saw is suddenly free.

London's Hornbeam Coppice

The team cheer. Sean, wasting no further time, drops down his helmet visor, restarts the chain and continues to sever the remaining limb with an undercut lower down. The colossal branch gives in and falls, taking its neighbour with it. On the ground both limbs are speedily sectioned into short manageable lengths and I join the volunteers in dragging the resulting brushwood to one side. We add these small branches to an expanding drift at the perimeter of the clearing and await the felling of another tree.

One often pictures chopping down a tree as a pretty straightforward motion. There is that neat image of the woodsman with an axe, hacking a wedge out of a straight and uniform

tree trunk before it submits to gravity and falls forwards in an orderly direction, leaving behind a clean, flat-topped stump close to the ground. Were the task to be so seamless in every case, however, trees would not be the characterful, individual creatures we so cherish and adore. When cutting down a mature broadleaf like an oak or chestnut, you are cutting, quite literally, though years of accumulated personality. Each season leaves its mark, often permanently, in the wood of a living tree, be it the pattern of its branches or the condition of its bark. Growth rings in the trunk, as everybody knows, detail the age of a specimen in years, but they also detail the nature of those years; an individuality that is retained in their appearance, distribution and thickness. The rings describe occasions of drought or fire, for example, through a narrowing or scarring of their lignified circles. They tell also of the direction in which the tree has faced – whether it has leaned towards the light or been forced one way or another by the weight of an adjacent tree – and at what point in its life this direction, or indeed *directions*, first occurred. And as each year's growth reacts to changing environmental stimuli, the resulting wood laid down commits these reactions to lignified, permanent memory. Only when the wood is cut are these experiences unlocked, proving to be an unpredictable medium that can spring and snap.

The coppice is situated within London's largest single area of woodland, Ruislip Wood, to the west of the city limits. That it takes a wood of this magnitude to gain the honour of 'London's largest' is a testament to the greenness of the capital. Its 726 acres (294ha), as Richard Hutton explains when we break for coffee, exceed the others in scale, including Queen's and Highgate Woods near Muswell Hill in North London. The entire landmass is listed as a National Nature Reserve, and protected by Natural England as a Site of Special Scientific Interest; it is a haven for a diverse range of wildlife running counter to the urban sprawl that partially surrounds it. Ruislip Wood is broken up into four main sections: Copse Wood, Bayhurst Wood, Mad Bess Wood and Park Wood, where Richard and his team are working today. As Woodland Officer Richard has spent over 15 years managing the enormous site, becoming well-acquainted, to say the least, with the many enchanting features of this ancient, semi-natural woodland. It's unlikely, he tells me, that he'll ever have the opportunity to manage a forest as big or as 'good' as this again; it is something of a dream remit when it comes to woodland management. The role encompasses what one might expect the job of a woodland officer to encompass: devising management plans, monitoring the wood's trees and habitats, maintaining its statutory paths, educating the public during summer months, and so on. But Richard's responsibilities also include the continuation of Ruislip Wood's heritage of coppice work, meaning that during the winter months he, together with another of the rangers, sets about cropping Ruislip's numerous stands of hornbeam, hazel and, to a lesser extent, sweet chestnut.

Coppicing is a sylvan practice most people are relatively aware of, such has been its resurgence in the realm of public attention. As the last decade has seen a revival of interest in both sustainable woodland management and the traditional

craft employed to deliver it, so has coppicing found its way onto prime time television programmes and national newspapers. To the uninitiated, however, coppicing – in simple terms – is the successive cropping of timber from the same tree at intervals. Some trees more than others respond to their main trunk being cut down by sending up a multitude of new shoots. These shoots in turn grow strong and tall, and can be harvested to produce lumber for a range of applications. The resulting base, shortened to a foot or so from the ground, is known as a 'stool', and will continue to live on happily enough as its top growth is systematically removed. Coppiced stools in fact will often far outlive a single-trunked tree; each harvest reinvigorating the stump as if beginning a new life. Hazel tends to be coppiced on a short cycle, producing small diameter poles that are both straight and long. These may be cut every 7–10 years, and put to use as garden stakes, peasticks, beanpoles and faggots (bundles of tied brushwood used for fire making and natural fencing). Chestnut is often harvested at longer intervals, allowing the wood to increase in size so that it may be used in fence making (e.g. post and rail), roofing (as rustic shingles) and as supports for commercially grown climbing plants like grape vines and hops. Chestnut coppices are therefore usually cut once every 15–30 years, depending on the employment for which they are intended. The wood of a hornbeam tree has many uses given its density and strength; it makes a very desirable material for construction and furniture, for example. But as a coppiced tree few contemporary applications are made using its poles, save for one – charcoal making – for which there is no better tree.

There has been coppicing in Ruislip Woods for centuries, and hornbeam stools – and indeed some chestnut and hazel – can be found in scattered spots throughout the vast nature reserve. Richard tells me a little of the shifting management of hornbeam throughout this period, how it would have once been harvested for ship building, before it adopted a more exclusive function as fuel for the growing city. Because of its density, charcoal produced from hornbeam burns hotter than any other native tree, and also makes great firewood. Coppicing at Ruislip ceased for a while following the First World War, when for a comparatively short period in the wood's five centuries of management the stools were neglected, growing broad and tall. As we walk through the coppice area in Park Wood we pass one of these relics, a multi-trunked brute, comprising five trees in one: a very beautiful tree. Richard indicates the original stool and where it was last cut, approximating the age of the current poles to about 70 years. This one will be coppiced, he tells me, along with the others at this site, but he will leave one central trunk to grow on as a standard tree. I ask him why: 'Just to keep some single, large hornbeams in the wood', appreciating that a hornbeam of such a size ought not to be entirely turned into charcoal.

In the early 1980s strategic coppicing in the wood resumed, its rotation in recent years coming under Richard's management. One of the benefits of restarting active coppicing in Ruislip is the diversity of habitats it has produced. The act of cutting trees down to their stumps, especially on a rotation scheme, means that intermittent glades are created in the wood, providing areas of greater sunlight that are favoured by certain butterflies, moths, birds and flowers. The extensive range of habitats was a contributing factor to the wood being designated as a National Nature Reserve in 1997, a status that remains in place provided the wood continues to be actively managed. Consequently, the coppicing at Ruislip is written into the ongoing management plan, just over 12 acres' worth (around 5ha), and on Monday and Tuesday mornings a little group of dedicated volunteers pitch up to assist with this programme. On most occasions, particularly during cold mornings like today, a shallow fire is kept burning beside the coppice stools, creating a central meeting place and restoring heat into cold hands. Nothing smells better to me than a midwinter wood fire, and the sight of one in the heart of an old forest like this produces a little rush of nostalgia. During my time as head gardener at a private garden in South London I would often keep a bonfire on the go, fuelling it with brushwood and – when working into the evening – baking the odd potato in its embers. The popping and cracking of combusting green wood is to me, as it is to so many people, a peaceful and calming sound, one I wish I could hear more often.

As employees of Ruislip Wood, Richard and Sean alone weald the chainsaws in its coppice. But the volunteers are clearly invaluable and well versed in processing the resulting lumber. Aside from the dispersal of the brushwood, all poles are sorted according to their quality and stacked in piles dotted around the coppice; I muck in with this task following their lead. When handling the hornbeam it amazes me just how much weight resides even in the smaller branches. In the past, every element of a hornbeam coppice would have been used, from large logs to the brushwood; even the leaves would have been gathered for use as mulch. Authentically, therefore, this kind of management practice was not particularly beneficial for wildlife. 'It's essentially intensively farmed woodland,' Richard tells me. 'What you're producing when you coppice is a monoculture, and that isn't brilliant for wildlife. Of course, there are a few butterflies and moths that are associated with coppicing, but most protozoa, invertebrates and fungi favour dark, damp woodland. The vast majority live under deadwood and in the leaf litter, so to remove it all is pretty destructive.' This isn't to say that all coppices are bad for wildlife per se; as already mentioned, the practice offers diversity of habitat within a woodland, and at Ruislip coppiced areas form only a small part of the overall makeup of the wood. However, Richard and his team make an active effort to leave a good proportion of the cut wood on the ground, being careful not to make things too tidy. The stacked wood

is divided, therefore, between log piles to be extracted later for processing into charcoal (which I will come to later), and log piles that serve the wood – for its creatures to inhabit and gradually return, as it rots, back to the soil.

Britain has taken a while to adjust to the idea that neatness in woodland does not represent wellness, and that abandoning a natural propensity to tidy is a positively good thing. The darlings of the wood, the much adored birds and mammals, benefit and indeed depend, in ecological terms, on dead wood. However, the creatures that relate directly to it tend to be those that are less appreciated. If we're being honest, insect larvae, beetles and worms do not inspire the same immediate empathy as those with fur or feathers. But the sooner a connection is made between the two, and of the dependence of one upon the other, the easier it is to see beauty in the decay.

There is a wonderful passage in John Stewart Collis's *The Wood*, whereupon leaving his ash copse Collis encounters a dead bird on the ground. For matters of general interest, he takes the bird home with him and places it in a basin in his shed to rot – just to see what happens. Collis describes the bird's speedy decomposition, and the creatures – the maggots, flies and worms – that bring it about. He visits the corpse often, witnessing each stage of the gruesome process, including the arrival of bluebottle maggots. Collis writes:

'I opened the flesh a bit more so that I might observe the main work of reconstruction. I gazed down at the tubes as they squirmed and twisted .... The bluebottle is necessary. The bluebottle is good. All things in Nature have a meaning and a purpose. All are necessary.'[1]

And then, as concluding justification for his macabre experiment:

'It is expedient, on occasion, to gaze down into the pit as well as up towards heaven, to look at the roots of Nature as well as at her flags, regarding the burden of the beginning and the dereliction of the end alike without flinching.'

He is right in bringing into question what we deem to be beautiful in the natural world. And where beauty fails, education can appreciate, for all living things belong to the same whole and are deserving of our interest. A woodland is, ultimately, one organism, and each element is as vital as any other. Thankfully, with the nation's largest conservation organizations promoting the merit and importance of dead wood, such as the Woodland Trust and the RSPB, hopefully it won't be too much longer before a cluttered wood is commonly valued.

Unlike the dead wood, Richard stacks hornbeam poles that are destined to become charcoal as close to Ruislip Wood's main paths as possible, so as to aid their eventual extraction. Other coppice workers have told me that cutting the wood is only one part of coppicing; extracting it from the woodland is every bit as arduous and delicate an operation. From a conservation point of view lugging heavy lumber across the soft earth of a forest can cause considerable damage to resident flora and fauna. Even if this was not such an issue the exercise remains a challenging one, particularly if the ground is still wet from winter, whereupon conditions become slippery and the soil is churned to mud. Furthermore, under these circumstances the use of four-wheel drive vehicles to perform the task can prove even more destructive. In this instance, and in those where a great deal of heavyweight wood is to be negotiated out of a dense forest, the traditional approach is often the best – the employment of a horse and trailer. Such a spectacle may appear old fashioned, but it is a remarkably efficient solution and still the preferred method undertaken by many coppice workers today. Whatever the chosen procedure, however, coppicing is always conducted with extraction firmly in mind. On this matter, Richard tells me that, given the not so enormous quota of coppice in Ruislip Wood's overall management, he tries, where possible, to restrict coppicing to easily accessible areas.

Once extracted – which Richard is able to accomplish using a small and relatively agile truck – the coppice poles are taken to another part of the wood and stored for a year while they season. Meanwhile the previous year's wood is chopped up into what is known as 'cordwood' (uniform sections of about an arm's length) before being stacked inside a charcoal kiln that Richard manages on site. I'd need to wait until spring if I wished to see the kiln in action, but it is a process I've been fortunate in witnessing on a couple of occasions in the past. Converting wood into usable charcoal using a traditional steel 'ring' kiln can take anywhere from 14–24 hours, and it is not uncommon for charcoal makers to leave a burn going through the night, checking on it at intervals. But the comparatively recent arrival of a more efficient kiln (pictured on page 50) – which is run in part by the cordwood's own gases being reused to heat the process – has decreased burn times to as little as eight hours. Whichever kiln is used, however, the concept is the same; moisture and gas are extracted from the lumber through intense heat, drying the wood without it fully combusting. Charcoal making is by no means a commercial enterprise for Ruislip Wood, but what proceeds are made return to the wood and contribute towards the continuation of its coppicing practice.

Towards the end of my time with Richard and the volunteers the sun pushes its way through the morning's cover of cloud, bathing the coppice in a brilliant wintery light. Stools dotted around the damp floor show white where they have been cleanly cut; both the sapwood and heartwood of hornbeam are particularly bright in colour. Modest sunbursts reach the woods beyond, creating patches here and there where the canopy has allowed light to break through. Next year, when this part of the wood is thick with new shoots and the seeding stems of the summer's willowherb and nettle, another parcel of coppice will have been felled, and another habitat created. Ruislip Wood is now in the throes of the earliest part of springtime, where a vitality in the air is tangible, irrespective of contrary weather. We've tipped over the point of mid-January prior to which all life is dormant; the days are now getting longer and the trees are slowly waking. Walking out onto the bridleway to begin the journey back home, I'm stopped by the wretched squeal of a vixen. I stand and listen quietly, and in the distant dark of the wood I spot a pair of foxes roll over one another and disappear into the trees. Less than five minutes later I'm back beneath street lamps and detached houses, and a patchwork of suburban roads into which I, in turn, disappear.

Autumn impresses no more characterful an icon upon the mind than a fallen, reddened maple leaf. Although it took the Canadians a century to commit its rudimentary shape to the centre of their flag, the maple leaf is readily associated with the wooded landscape of their southern provinces. In particular, the deciduous, broadleaf woodland stretching down through the lower reaches of Ontario remains one of the world's centre points for captivating autumnal colour. With roots dug firmly into the foundation holes of a once ancient woodland, the tree species composition here remains more or less as it has been since the Ice Age: a familiar mix of temperate zone regulars such as poplar, ash and red oak (though the absence of American chestnut, all but eradicated by an imported blight early in the 20th century, marks a noteworthy alteration). Follow the Niagara River just a short way upstream from its famous falls and you'll find yourself immersed in the spectacular surrounding of the Great Carolinian Forest, a surviving pocket of virgin woodland.

At Niagara Glen Nature Reserve, to reach the wood itself one must drop down into the glacial river valley, a descent through vivid leaves onto a floor of spongy, leaf-sprung soil. This is the second of two autumn visits I am making to the Carolinian Forest. The first was to Morpeth, on the southern tip of Ontario, separated from America by the Detroit River. The river flows into Lake Erie, circling Morpeth's peninsula of Rondeau park, a protected area of natural woodland. The reserves of Rondeau and Niagara Glen are known to share between them the truest representation of undisturbed, original Carolinian Forest. Having long wished to experience the changing season in this magnificent topographical relic, I've made my way to these two particular locations of conserved natural beauty, just as the leaves are turning. Despite its distinguished charm, however, it isn't only the maple leaf I've come to see. These woods along the Canadian–American border are home to another autumnal champion, the ironwood.

Ironwoods of Ontario

The ironwood, or 'musclewood' tree is the American hornbeam, *Carpinus caroliniana*. So-named on account of their robust, fluted trunks, ironwoods have a long history of pre-industrial utility. In the days before the commercial availability of steel, timber from these stout little trees was traditionally appropriated for the construction of water pipes, windmill wheels and butcher's blocks. In fact, the wood grain is so dense that it provided the central crosspiece once fitted to the shoulders of an ox, bracing the load of heavy farming ploughs as they were towed through the ground. As one of the Carolinian's chief representatives, the turning, serrated foliage of the American hornbeam contributes an explosive golden-yellow to the dramatic kaleidoscope of vibrant colour, making it stand out in the lower storeys of the woods. There are more brilliant and brighter yellows in the leaves of Niagara's trees – the strangely shaped *Liriodendron tulipifera* or tulip tree, for example, or those of the poplars – however, the orange tint is a detail in this environment unique to ironwood foliage.

Carpinus branches have a distinct nature of their own too: a low, spreading habit forming horizontal veils that drape below the forest canopy. It doesn't take long before I find myself looking up through the branched layers of an ironwood, standing in the quiet of the wood. As its leaves catch the sun I can see the dried, hardened veins running out from their leaf margins. The tree I'm looking at is probably only a little over twice my height, however, it is already developing the splayed habit idiosyncratic of its woodland form.

Hornbeams planted in municipal parks are often of a different habit entirely. Like many ornamental trees given, you could say, an unnaturally wide berth in a park setting, hornbeams display a more compact form uncharacteristic of their forested ancestry. The shape is fuller and the leaf coverage denser, forming an impenetrable profile supported by a framework of rigid limbs. It's a stunning shape in isolation, and the architectural embodiment of a classic tree form, but hornbeams seen in this context lack the wonderfully eerie quality of their wooded relatives. The true hornbeam is a tree of the forest, its figure brought to life by the reach of its muscly, sun-striving branches.

However, this stretching habit in the lower storeys of the woods cannot be divorced entirely from the shapely form seen in parks and gardens. As one of the shorter trees of its native forests, it could be surmised that the ironwood is

exposed to grazing in a way that other trees are not, leading to a high tolerance for being 'hacked back' as leaves and small branches are consumed. This tolerance therefore lends itself well to being clipped, which is why hornbeams are so well-suited for hedging and ornamental shaping with shears (as well as more brutal arboricultural equipment).

One may rightly assume from the name that the Great Carolinian Forest is not exclusive to Canada. Its virgin footprint does in fact stretch south from Ontario down to the states of North and South Carolina. While only patches remain of the original coverage, an outline of the forest can be drawn in an arch between North and South, taking in eastern states such as the Virginias, Tennessee, Pennsylvania, New York and Ohio.

The Niagara river, muffled up until now by a cloak of falling leaves, comes into earshot as I emerge through the trees. The vast body of water cuts a giant path through the woods, dividing two countries with a surging current. However, the woods either side are of the same constitution. It is only from the riverbank, in fact, that I'm able to take in the extensive radiance of the Carolinian Forest as one. Divisions like this, such as a road through a wood, offer an advantage in perspective otherwise hindered by arboreal congestion. John Stewart Collis wrote of the overuse of the phrase, 'not being able to see the wood for the trees'. He notes: 'it might quite easily mean nothing and yet be repeated twice daily by our publicists.' But looking across the Niagara from one bank to its opposite, there comes a 'vision of the whole'. As Collis puts it, 'a too careful dwelling upon many particulars blinds us ... you cannot catch sight of the wood as a totality if entangled in the trees.' The sight in this case is a distinct set of colours, and it is now possible to further observe a multitude of distinct hues. Comprising a range of reds, oranges, yellows and greens the overall Carolinian coloration is mesmerizing, and it takes a while for me once again to unpick its components. There is the tall, majestic form of tulip tree, and then the rotund, stockier blotches of walnut, chestnut and beech. There are the faithful reds of maples and flashes of bright red dogwood, and, woven in, thuja greens and those of fading oak. In no other season may a woodland's constituent trees be made out in this way, distinguished by a subtle contrast of colour. Beneath the splendid canvas I can make out the ironwoods, their venturing limbs just about perforating the canopy, glowing like fire from within the forest.

Douglas-fir

Distinct from the true firs (*Abies*), the common and botanical names of the Douglas-fir reveal much about the tree. Its generic name is a combination of two words: 'pseudo' meaning 'false', and 'tsuga' from the Japanese name for a similar genus of conifers; the common name of hemlock given to species from this genus is derived from an apparent similarity in smell to the crushed leaves of the poisonous hemlock plant. The species name, *menziesii*, derives from Archibald Menzies (1754–1842), the Scottish physician and plant collector on HMS *Discovery*, who first collected a specimen of the tree. The common name extols the name of another Scot, botanical explorer David Douglas (1799–1834), who brought back the seeds on a subsequent voyage to the same North American region, and introduced the tree to Britain, where it has played a significant role in plantation forestry.

There are only a handful of species in *Pseudotsuga*, and two varieties of the Douglas-fir itself. This chapter is concerned with the coastal variety, one of the world's enormous conifers – a tree of immense stature and heritage. However, there is also an inland variety, the Rocky Mountain Douglas-fir (*P. menziesii* var. *glauca*), which is native in the USA with a range as far east as Wyoming and south to New Mexico. While deciphering the identity of the many similar-looking conifers can often be a challenge, a citrusy scent distinguishes Douglas-fir foliage when crushed.

Douglas-fir (*Pseudotsuga menziesii*)

Tony's cabin is a short vertical walk up onto the hillside above his home. He has suggested that I go and see it as the view describes the contours of his land better than any other on the property. Climbing up through a bracken-laden field I reach the cabin and sit on the edge of its levelled platform, boots dangling over the planks into a clump of nettles below. The modest wooden structure looks across Tony's stretch of narrow valley: its sharp drop leading down to a slip of stream and then climbing back up through thick forest on the opposite side. The panorama is something quite special. It offers a sense of seclusion unlike any I can recall experiencing in Britain. In this spot, with the hill continuing its rise behind me and the entirety of the view in front, I feel at once protected and exposed, shrouded in a landscape devoid of modern construction. It is the type of scene immortalized in the imaginations of city-bound nature lovers; the kind that is almost impossible to find under circumstances unblemished by tourism.

As I sit, my eye drifts from one green hue to another, making out the shapes that dictate their variation. The bracken in front is light with midsummer growth while the grass, mown close by a ranging herd of red deer, tints yellow like moss. The sky is overcast and bolsters a light breeze on which swallows dart, swooping past the cabin and into the valley below. I watch the deer for a while, moving slowly in and out of view, grazing silently in a loose but loyal group. And then all of a sudden, cutting right through the serenity, comes an electrifying wail. It carries long and mournful up from the trees and out over the valley. The deer are alert and upright, their necks stiff and tight. I'm aware of a similar sensation in my own shoulders; I've never, in the flesh, heard the cry of a wolf before and my reaction surprises me. I might have expected it, as the wolves are what brought me here, but such howls adopt an unanticipated intensity when separated from physical form; there's no sign of a wolf, nor is it easy to pinpoint exactly which part of the wood the call has come from. The trees ahead are so dense with evergreen foliage that the sound is impossible to trace. The mass of tall trees now looms with new vehemence, concealing their elusive occupants under an impenetrable green thatch.

Plantation Wolves

The cry will have come from either Sansa or Rickon, two relatively new arrivals to Tony Haighway's Shropshire conifer plantation. Tony purchased this wood during the developmental years of what is now Wolf Watch UK, a rescue centre dedicated to wolf conservancy and care. His 70-acre (28.3ha) sanctuary – nestled discreetly within the wild and rugged borderland between England and Wales – plays host to wolves of varying age and species, most often casualties of zoo closures, dominance fights and overbreeding. The organization is listed as managing some of the largest wolf enclosures in Europe – an unlikely statistic to be attributed to the landscape of Great Britain. However, access to the sanctuary is far more limited than for zoos and wildlife parks, and is offered only to membership subscribers through pre-booked arrangement. Tony's ethos is that the wolves come first: their privacy is paramount and any disruption to their occupancy minimal. To this end the sanctuary's setting is as close to the wolves' natural habitat as possible.

It was the setting that first interested me; how does one accommodate this kind of animal? What British terrain could provide wolves with a familiar, hospitable environment? It's been centuries since the European wolf was a wild resident of the UK, and even longer since it roamed free from human persecution. Britain's landscape has altered dramatically during that time, to the extent that picturing such an animal on our land, however green and pleasant, verges now on the fantastical. Assuming there had to be some form of woodland involved, I contacted Tony on the basis that I was writing a book about trees and was interested to know what species, if any, form the body of his wolf enclosures. Tony obliged with an answer that in hindsight I ought to have expected, 'Ex-forestry,' he said, 'mostly Douglas-fir.' As such, the woods at Wolf Watch UK are more alien to the landscape than the wolves themselves – non-native softwoods introduced from North America. They are exactly the kind of trees one might associate with wolves; those we picture residing on the slopes of snowy north-western mountains.

During our phone conversation Tony outlined how he'd acquired the wood through fortunate timing and sheer luck. Wolf Watch began at his rural home in Warwickshire, but by the early 1990s, interest from zoos and wildlife centres in rehoming wolves with Tony had led to the need for expansion, and so he began looking to relocate the sanctuary. 'I drew a circle around Warwickshire of 100-odd miles, tripped over into Shropshire and found this lovely place tucked in the middle of nowhere,' he told me. The location ticked all the boxes; secluded yet accessible, nestled in a private valley thick with mature plantation woodland. The property came with 17 acres (6.9ha) attached, but Tony quickly expressed interest in purchasing some more of the surrounding forest from its proprietor. Sadly, at that time the asking price was too high, but a few years later he learned that the forest was due to be clear-felled 'from top to bottom' and sold for timber. 'That would have absolutely wrecked the whole environment and ecosystem that existed within it.' Tony went to building societies and banks to find out how much money he could borrow, made an offer for the land and to his amazement it was accepted. The sudden affordability resulted from a huge storm in France that had recently devastated European forestry plantations. 'Hundreds of thousands of hectares of softwood trees had hit the deck,' explained Tony, 'saturating the French and wider European market with available timber and having the knock-on effect of devaluing softwood forestry.' With the trees purchased and the forest secured, he began moving his wolves into their new and expansive home; a home well suited to the requirements of their species. Excited to hear about this unconventional use of plantation forestry, I arranged to visit Tony at the sanctuary and to meet some of the wolves; to witness a once uniform forest set loose as a wild refuge.

When I return from the cabin Tony brews coffee over which I relay my excitement at hearing the calls in the woods. The farmhouse kitchen is lowly lit and homely, warmed by an Aga at the base of which sprawl a deerhound and springer spaniel. Tony details some of the interest Wolf Watch UK has attracted over the years – documentary teams, including the BBC, have filmed the wolves for a number of programmes. He considers each request according to what best serves the wolves, granting access to those predominantly of an educational agenda, provided that the animals remain essentially undisturbed. Tony has little interest in the notion of 'reintroducing' wolves to the UK, a sensationalized topic on which he is often incorrectly assumed to be in favour. Through the kitchen windows the sky has begun to brighten, and we decide to take advantage of the sunshine and make our way out to the enclosures.

Our route follows a stone track that leads from the house up into the trees, passing the little cottage that will be my home for the night. We enter the edge of the wood at the base of the valley and climb steadily upwards into progressively dimmer light. On both sides of the steep track run enormous towering fences. The narrow road into the wood divides two large wolf enclosures, forming a central service channel large enough for a single vehicle. A short way up we reach a vast gate that Tony opens with a great clang, the abruptness of which is dulled by the surrounding trees. The track ahead takes us deeper into the wood and among the firs themselves.

The rich green appearance of Wolf Watch's Douglas-firs is, from the outside, very much at odds with the colour of its interior. You might not expect it to feel so cavernous and open for one thing: save for the tall, immensely straight trunks there's very little to obstruct long vistas through the trees. There's also comparatively little greenery below the canopy – the vast majority of side shoots having dropped away on account of the dwindling light. This leaves the trunks and split branches a dusty brown colour, the kind of dishevelled look one might mistakenly attribute to ill health or poor upkeep. However, this all fits neatly into what is expected of a plantation forest: few other trees are able to take hold in the shade, and this means that the firs have little to compete with besides one another. And while competition leads to a slow de-branching in the lower storeys of the wood, entire trees also fall prey to the tight crowding, with the victims lying at odd angles throughout the plantation or remaining 'hung up' in the branches of their neighbours.

Tony describes how leaving fallen trees to break down naturally has obvious benefits for this resident wildlife, not least in providing additional habitats to accommodate the diverse inhabitants. He's happy to manage the wood in this fashion, 'provided this doesn't compromise its fundamental existence in any way.' He relates a desire for the forest to remain untouched; an organic environment left – where feasible – to its own devices. This perspective echoed a favourite passage of mine by nature writer Roger Deakin, a diary entry transcribed in *Notes from Walnut Tree Farm*:

'I can't bear to mow my lawn because it would mean mowing all the blueness out of it, the vanishing blues of self-heal, bugle and germander speedwell. They are worth more to me than the neatness of a mown lawn.'[1]

Wandering close to the enclosure fence line we become aware of something observing *us*, and turn to see a motionless wolf fixing a gaze through the trees. She's probably been watching the whole time, I say, 'Almost undoubtedly,' replies Tony, happy to see the wolf coming so close. 'She's becoming familiar with her surroundings. A couple of weeks ago she would have been off and away, but bit by bit she's realizing that we're not a threat.' He calls her name, 'Sansa', softly back to her to affirm the inquisitiveness. Sansa is a Norwegian wolf, sleek and Alsatian-like (to my uneducated eye), more slender than my preconceived image of a wolf. Within less than a minute though she's trotting away back down the steep hillside, negotiating its gradient with a steady, hereditary ease. Sansa has only been at the rescue centre a few weeks, but already she appears confident and familiar with the terrain of her new enclosure. The wolves, however, are not alone in their ease with the slope of this landscape; this is in fact another commonality shared between wolf and tree.

Douglas-fir plantations are a common sight on the steep hillsides of England, Wales and Scotland; widespread on suitable ground, usually grown for commercial timber. But, as mentioned above, their presence on British soil has a comparatively short history. It was only in the late 18th century that Archibald Menzies stumbled upon the tree in western Canada, and it was some years later, in the 1820s, that its seed was fastidiously collected and sent to Britain by Scottish botanist David Douglas. The introduction of the Douglas-fir (as well as other significant conifers, shrubs and flowers) is a tale of relentless and admirable devotion, regularly at the expense of comfort and health. Douglas's travel diaries, thankfully preserved and printed, read like a real-life Robinson Crusoe, wrought with narrow escapes and pragmatic self-sufficiency. They detail a comprehensive list of the plants encountered during his trips, recounting the circumstances and conditions in which they were found growing. In years to come these meticulous observations would prove vital, particularly when establishing a new flora in foreign, European soil.

'He was a Scotsman made good,' laughs Syd House, Douglas biographer and retired Conservator of Forests for the Forestry Commission, Scotland and, like Douglas, a Perthshire native. 'He was of humble origins, son of a stone mason and an apprentice gardener.' Speaking on the phone sometime after my visit to Wolf Watch I reach him at home in south-central Scotland. 'Douglas was fortunate in working under William Hooker at Glasgow University Botanic Gardens,' says Syd. 'Hooker later recommended him to the Royal Horticultural Society (RHS) as a potential plant collector. He was used to walking, used to the Scottish hills, and educated so that he could write and record. By this time, he had clearly become a very skilled naturalist.' Douglas's fastidious recording of site and soil was central to the success of his plant introductions to the UK. In Northern America and Canada, he observed colonies of Douglas-fir populating steep mountainous slopes, growing in close proximity and always straight upwards. 'The firs displayed negative geotropism,' says Syd, 'meaning that they naturally grow away from the direction of gravity.' The trees were therefore able to traverse this rising, dramatic environment while remaining perfectly straight and upright.

Negative geotropism comes in pretty handy when you want to produce tall, straight timber in large volumes. Britain's demand for fast-growing, linear timber had never been so prevalent as in the years following the Great War. David Lloyd George, Chancellor of the Exchequer during the First World War, is famously quoted as saying that Britain '... had more nearly lost the war for want of timber than of anything else.' Woodland resources had been so plundered over centuries of industrial development, and furthermore during the recent war effort, that Britain's total tree cover was at an historically meagre 5 per cent. The war itself had also made European timber imports increasingly uncertain, mounting pressure on the country to regain self-sufficiency. It was decided that refortifying the UK with its own forestry resources was of the utmost priority, and in September 1919 the Forestry Commission was born – a new governmental department charged with replenishing timber reserves.

Of the trees planted during this great push for afforestation, one of the key players was the Douglas-fir. During his explorations, David Douglas had recognized that growing conditions in the Pacific Northwest had similarities with western Europe; both areas were on the edge of continental landmasses, and both were significantly influenced by maritime weather conditions. Douglas's skill as a plantsman led him to surmise that the Douglas-fir (among the numerous other conifers he collected and sent home) had great potential as a forestry tree, long before it was put to use as such. By the time the Forestry Commission was established therefore, the Douglas-fir had undergone almost 100 years of testing on British soil. Britain has a long history of introducing trees owing to its relatively poor range of arboricultural flora. Syd House tells me that this is predominantly a consequence of our separation from the European continent. 'The English Channel put a stop to trees recolonizing following glaciation, which is why we have so few native conifers compared with mainland Europe.' Plants brought back by the explorers of the 18th and 19th centuries were cared for and propagated by botanical institutions such as Kew Gardens and the RHS. But they also found themselves planted into the gardens of wealthy landowners whose patronage had subsidized the expeditions. Indeed, at the time of writing, the tallest Douglas-fir in Britain is over 217ft (66m) high – the tallest tree in Britain in fact – and can be found near Inverness in Scotland. It stands in a grove of firs planted on what was at that time a private family estate.

Proving particularly successful in less favourable conditions was what made the Douglas-fir so attractive to the Forestry Commission. Purchasing land for the expansion of forestry was an expensive business, and by the 1950s the Commission soon began looking to upland sights as sources of cheaper realty. It became necessary, therefore, to employ trees better suited to such terrain, those that thrived naturally on steep slopes and deep valleys. By the end of the 20th century Douglas-firs populated nearly 47,000 acres (19,000ha) of Forestry Commission land.

There were, of course, drawbacks to this large-scale introduction of non-native conifers. For one, the monocultures they created were deemed to have negative environmental effects, such as altering soil pH and restricting biodiversity. Their dense canopy inhibited light from reaching the forest floor, resulting in cooler temperatures and thus poorer breakdown of organic material which would otherwise return nutrients to the soil. These issues were widely reported on and have subsequently led to changes in the official approach to commercial forestry. Invariably, a wider range of tree species will offer more for the faunal diversity of a wood than any single species planting, and consequently the UK's forestry standard now advocates that no more than 75 per cent of a given plantation may be dominated by a single species of tree. Irrespective of this, mature plantation stands still have a great deal to offer. The pages of Neil Ansell's book, *Deep Country*, for example, act as a voluminous account of the wealth of life thriving in the spruce monoculture of his beloved 'Penlan Wood' in mid-Wales. This is also backed up by the long list of resident birds and mammals Tony Haighway recounts as having been seen in the woods at Wolf Watch. And, after all, what other forest in Britain is home to wolves?

Back at the sanctuary, later the same day, Tony introduces me to a large Canadian wolf named Madadh, the most senior of the wolves at Wolf Watch. Madadh's sociability also sets her aside from the others, having been reared by Tony as a very young pup. She has her own enclosure, close to the house, which is laid out around the stream at the bottom of the valley. It's less wooded in this area, although not without trees, but is comfortably suited to Madadh's increasing frailty. The enclosure offers her a manageable range of environments, from tall grass meadowland to a shallow pond. As we pass through another large gate and into her space I feel a rush of excitement. She approaches quickly, inquisitive and confident. When she brushes past me for the first time I'm aware of an innate strength, the weight and force of which is palpable, despite her old age. Up close, wolves really are remarkable animals; familiar yet wild, unequivocally magnificent. I watch as Madadh trots to and from us, returning frequently to Tony, clearly fond of his presence. 'The only way you get a strong bond with a wolf is if it comes to you somewhere in those first few days,' he says, 'just as its eyes are opening. It's been recognized that that initial three-month period is the crucial bonding time where socialization can take place.' The resulting close relationship he now shares with Madadh does not exist with the other wolves at Wolf Watch, at least not in the same way; each dynamic is different, developed patiently over years of gentle interaction. Like the many creatures in Tony's wood, the wolves are free to respond or, in turn, not respond to his presence, free to treat the forest as home and habitat.

Meanwhile the forest, liberated from commercial felling, continues to evolve with the wolves, similarly encouraged to remember its roots. At the time of writing, the tallest Douglas-fir in the world lives in Western Oregon, USA. Measuring close to 328ft (100m) in height it is almost one-third taller than the fir at Inverness, although it is also thought to be twice as old. What isn't to say that the giants of western America and Canada may one day be walking as tall in Britain as they do in their native homelands? While the vast majority of Britain's Douglas-firs, noble firs, Sitka spruces and western hemlocks will continue to be felled for planks, replanted and then felled again, plantations like Tony's – those put out to pasture if you like – may live on to command the landscape as Goliaths among the native flora.

As night falls at Wolf Watch I retire to the comfort of the guest cottage, a neatly converted granary accommodating visitors to the centre. It's a sweet little place with two bedrooms, a kitchen and upstairs living room, the latter brimming with books of a kind conducive to a weekend with wolves. The shelves carry everything from natural history to modern fiction thrillers; all stories of wild nature to accompany guests during their stay. Phone reception in the valley is all but absent: no better escape from the demands of modern communication. I make a mug of tea in the wonderful silence that has settled around the cottage and take it downstairs along with a copy of Mark Cocker's *Crow Country*. Opening the bedroom window, I fix it ajar with the catch, hopeful that I might be awoken – suddenly in the night – by a howl from the feral forest.

Firs on
Vancouver Island

Today is 21 August 2017 and America awaits the arrival of its first solar eclipse in 38 years. The last time an eclipse of this magnitude passed fully between the Pacific and Atlantic coasts was in 1918, and in anticipation of its recurrence the news channels have reported on little else for the last week. At Los Angeles airport the outdoor terrace of Terminal 6 is a hive of activity: passengers cram onto the narrow deck talking excitedly and passing around makeshift protective sunglasses. LA is too far south to be graced by the impending 'path of darkness'; however, at any moment the eclipse will come partially into view – reason enough, it seems, to gaze hopefully at the skies. Waiting for my flight north to Vancouver I look up a digital projection of the shadow's course on my phone. The arcing path bears a striking resemblance to the Oregon Trail of the 1800s, only in reverse; it traces the emigrant wagon route through the Midwest to Missouri, before vanishing at the beaches of South Carolina in the east. For many, solar eclipses belong to that particular category of natural events so wholly beyond human influence as to provoke a certain emotion in their observers. It is a kind of thrill, I suppose – especially in the modern age – a reminder of the magnitude and power of the Earth in which we are otherwise so much at home. I've experienced it in the obvious places, such as Niagara Falls, where the sheer force and volume of water causes the stomach to suddenly plunge and the ground to appear unsteady. I've felt it also when caught out by a passing storm in Iowa; my stationary car shaken violently by fierce wind and rain. It is a skipping heartbeat in the presence of something primitive and autonomic. Strangely though, matters of astronomy have never quite succeeded in manifesting with the same physicality for me. Today I'm content in allowing the great shadow of America's 2017 eclipse to pass me by as my attention lies elsewhere. I'm travelling to the Pacific Northwest to witness another of the world's natural marvels – another wonder, in fact, casting a great shadow across the land.

Following a night in Vancouver I catch an early ferry over to Nanaimo on Vancouver Island. The view is irrepressibly spectacular – a vast plateau of water framed by the mountains of British Columbia, which rise on all sides from its splintered western isles. During the journey across I try and picture what it must have been like to sail to the Pacific Northwest as a plant collector in the 19th century, as David Douglas did *en route* to discovering the giant conifers of the region. Amazingly it seems no significant development yet hampers the panorama; the setting here appears fresh, for want of a better word, cosmetically unspoiled in a way landscapes rarely are. The view remains, quite simply, green of land and blue of water.

On arrival I continue northwards along the island a short while to Parksville, a small city on the east coast. The area is home to stretches of coastal rainforest, the interiors of which contain one of the world's grandest tree species, the Douglas-fir. Familiar with this tree as a prominent constituent of timber plantations back in Britain, I've come to experience its ancient form growing here in the forests of British Columbia. It is often the case that these conifers are, quite literally, overshadowed in popularity by the sequoias further south, which are a number of feet taller. Fascination with the continent's 'big trees' tends to divert visitors towards Yosemite and Sequoia National Parks, but unlike the redwoods, Douglas-firs thrive in a temperate climate more relatable for travelling Brits. Along the Pacific coastline they pick up where the former drifts off, sharing ground at the top of California before dominating the Oregon and Washington skylines. In this region of cooler weather, the Douglas-fir has thrived for thousands of years, and on Vancouver Island, next along this northern progression, some of the oldest of its species are found in remarkable groves; pockets of ancient, old-growth forest that retain a unique and precious ecology. Magnificent Douglas-firs can be found in numerous locations on the island, although sadly what remains of the original forests is thought to make up less than 1 per cent. Since European colonization in the 19th century, unregulated logging has devastated Vancouver Island's natural resource, isolating its old growth to patches scattered among adulterated forest.

Close to my rented Parksville apartment there are two Douglas-fir groves I have made arrangements to visit. The first is Cathedral Grove, a popular destination among visitors to the island, colloquially dubbed for the enormity of its tree trunks and the shafts of celestial light that filter through the canopy. The second is a less trodden stretch of forest at Qualicum Beach a few miles north, cared for by the local community as an ecological reserve. In the morning I'll be meeting Gary and Ronda Murdock there for a 'forest bathing experience' – an alternative introduction to the ancient forest trees. The Murdocks are Vancouver Island residents, Gary a former forest technician and Ronda something of an ethnobotanist. Together they share a great love of the island's native history and flora, guiding nature hikes through the local rainforests and helping preserve their ancient heritage. Forest bathing is a meditative practice that has found its way to British Columbia from Japan in recent years. It originated as *shinrin-yoku*, a Japanese therapy centred around quiet contemplation in the presence of trees. The concept is nothing groundbreaking – it is simply a walk in the woods – but the results recorded by researchers in Japan ascribed a lowering of blood pressure, heart rate and stress levels, suggesting *shinrin-yoku* to be an altogether restorative activity. I've written about forest bathing in articles before, equally in favour of the concept as I am questioning of its medical effects. However, I hadn't yet experienced the practice first hand, so when I stumbled upon Gary and Ronda's rainforest walking tours I figured this would be an opportunity to try it for myself. Rather than Cathedral Grove Ronda suggested that we meet at Qualicum Beach, as it would be comparatively quiet, and more conducive to a contemplative amble.

Settled into Parksville, and with a few hours of the afternoon still remaining, I head inland to Macmillan Provincial Park, eager to acquaint myself with the firs of Cathedral Grove. The short journey takes in Cameron Lake, a narrow slip of beautiful blue water enclosed by stony beaches and spreading forest. White sprays of holodiscus protrude out from its banks like tall, blowsy meadowsweet, and yellow heads of tansy clump here and there at their feet. Towards the end of the lake the road diverges, exchanging the open water view for the diminished light of mixed forest. As I go deeper the deciduous leaves of spring maple and red alder gradually dissolve into a greater volume of evergreen needles; tree trunks become wider and lower-storey branches thin out. A little way further along I realize I'm now in the company of hemlock and western red cedars, ecoregion partner plants associated with coastal Douglas-fir. I pull off the road and join a handful of cars in a dusty car park. A boardwalk leads off into the wood itself, now somewhat darkened by the onset of evening. At the start of the boardwalk I read a notice sign reminding visitors that the forest is very old and therefore in a state of undeterminable flux; branches will 'fall without warning', it states, and it is better not to enter at all on windy days.

The Scottish adventurer John Muir once wrote a dramatic essay wonderfully undermining the latter principle of this warning sign. The first-person account, written in 1874, documents his experience climbing an enormous Douglas-fir tree during a ferocious storm. The essay was published in *Scribner's Monthly* – a New York-based periodical – during Muir's years exploring the wild terrain of California's Sierra mountains. His account describes the event as both 'beautiful' and 'exhilarating': '... when the storm began to sound,' he writes, 'I lost no time in pushing out into the woods to enjoy it.'[2]

Why John Muir would choose to experience one of America's 'most bracing wind-storms conceivable' from the hazardous environment of a harried forest is answerable only by his consistency to indulge such perilous whims. As the storm rages through the Sierra, Muir locates a group of what he noted as 'Douglas spruces', and climbs the swaying trunk of the tallest among them. From this elevated position he records the storm's interaction with the forest, recalling the distinct and individual manners in which trees respond to the force of the gale. The 100-ft (30.5m) Douglas spruce to which Muir clung 'like a bobolink on a reed' is now understood to have been a Douglas-fir. (Given the comparatively recent discovery, let alone the nomenclatural dedication of the great wealth of conifers found in this western region, his mislabelling ought to be overlooked.) 'The Douglas spruces,' he writes, 'with long sprays drawn out in level tresses, and needles massed in a gray, shimmering glow, presented a most striking appearance as they stood in bold relief along the hilltops.' This is the imagery that brought me to Vancouver Island: a western landscape similarly celebrated for its impressive and commanding firs.

A short way along the boardwalk reveals a multitude of mature Douglas-fir boles. Aside from the enormous girth of their perfectly straight trunks, the trees are readily identifiable by the density of their bark. This heavy, cork-like bark makes Douglas-fir particularly resilient to forest fires, forming a protective buffer around the core. Cathedral Grove, like many forests in the Pacific Northwest, is known to have been besieged by centuries of wild fires, the memories of which are often retained by visible scarring on its trees. Interestingly, as the thicker bark only develops in mature Douglas trees, it is these older, taller specimens that survive such forest fires. Taller firs also have the benefit of fewer lower-storey branches, preventing fire from reaching the combustible needles. Over 300 years ago a fire at Cathedral Grove knocked out many of its trees while the older Douglas-firs remained standing as they do today. These veterans form the footprint of the ancient grove, some of them thought to be at least 800 years old.

When encountering trees like these for the first time there is an impulse to approach and touch them; an irresistible interaction that makes children of us all. I place my palm on the dusty exterior of an enormous fir. Its density is palpable; a solid girder in the ground. The tree seems at once inanimate and alive, and to touch it excites a reaction of fleeting perplexity. What I can feel isn't so much an 'energy' but a tangible sense of endurance; of immense dead weight bound up in a living, breathing organism. The effect is momentarily humbling. The second impulse is, of course, to look directly upwards. The spectacle is appropriately dizzying; it is a huge tree – in all likelihood over 200ft (61m) tall – and as straight as a pencil. The trunk remains branchless for a great distance, and when limbs do appear they're stout and relatively congested, like a conical bottle brush. Way up there I can just make out the wind; a soft, continuous fizzing like the sound of a distant beach. I spend a happy hour wandering through Cathedral Grove, acquainting myself with its distinctive yet unfamiliar flora in the fading daylight. The forest floor is a mix of ferns and sapling trees, and plants like the curious 'Devil's Club' bearing little thorns on its palmate leaves.

The next morning I meet Ronda and Gary Murdock in a car park outside town. Following introductions, we drive northwards along the coastline to Qualicum Beach, passing swathes of madrona trees (*Arbutus menziesii*) with their striking red bark. These are another tree iconic of the region, named in memory of the same botanist who discovered the Douglas-fir, Archibald Menzies. In the car Ronda, brimming with local knowledge, gives an enthusiastic overview of the Island's history. She introduces the forest we're headed to and its timely preservation, noting the indigenous peoples that once occupied it and the British Army general, Noel Money, whose trading company purchased the 55 acres (22.3ha) it stands on. Following Money's death in 1941, the plot – which, in addition to the forest, included a mansion house hotel and substantial golf course – was bought by an affluent local family and remained a private estate until the late 1990s. Around this time residents of Qualicum Beach, familiar with the estate and fond of its old-growth forest, got wind of plans for the plot to be sold once again. With stands of ancient Douglas-fir already so heavily in decline on the island, and no government law as such preventing their destruction, the local community took the initiative to form a preservation society and bid for the land. Granted the time necessary to raise the significant funds (in excess of a staggering $2 million), the group were able to purchase the forest and dedicate it as a heritage site open to the public. That such a magnificent stretch of endangered rainforest exists so close to a town is testament to the significance of their achievement.

The wood at Qualicum Beach is bathed in mid-morning sunshine and here and there the ground blazes with concentrated radiance. Assembled in an inward facing circle only a short distance from the woodland entrance, our little group begins its meditation. My ears take time to adjust in the quiet, first humming with the familiar tinnitus of travel and sleep disruption, and then gradually begin to discern individual surrounding noises. The wind high above becomes clearer and softer and occasional cars are faintly audible on the road outside. Ronda allows a few minutes to pass before speaking softly over the quiet. 'We are grateful for this forest,' she pronounces, turning to Gary and me, 'feel your body; your limbs, the cool air. Feel your legs and feet on the forest floor.' Ronda asks us to imagine ourselves possessing roots; roots that lead down from our feet into the ground below. In this prelude to the walk attention is drawn to our own physicality within the forest, so that the walk itself can therefore be occupied with the physicality of the forest: its temperature, sounds, sights and smells.

A little bell chimes; we open our eyes, and Ronda announces that we will now begin the walk. 'During this first section I want you to turn your attention to the small things; look closely at whatever holds your attention and study it a while.' A part of me is self-aware and conscious of our appearance to any bemused onlookers, but I will say this at the outset – though it may seem like an obvious statement – no activity is as peaceful as drifting aimlessly through a sunlit Canadian rainforest in the company of the horticultural equivalent of a yoga instructor. Everyone should try it. Following Ronda's direction, I focus initially on the small things, overriding an internal urge to first identify the surrounding plants. There is minuteness all around: a caterpillar suspended on invisible thread; amber-coloured resin oozing between strips of bark; berries on the low Oregon grape (*Mahonia nervosa*) that litter the forest floor. After ten minutes or so Ronda's bell chimes again. We regroup beneath a pair of huge Sitka spruce and take it in turns to discuss our observations. We stand and listen for a while before moving to the next focus, 'motion'. 'If a movement in particular takes your attention, go with that,' Ronda says, adding that while some things may appear at first to be still, we should take a few moments to see whether they are in fact moving in some way.

This narrowing of attention from one element to another seems to me the art of forest bathing and the reason for its calming results. When discussing examples of motion, it seems we have found it in all manner of places, from light travelling down spider silk, to dancing fronds of bracken. Similarly, as we turn to the sense of touch, I find myself drawn to differing patterns of bark: long strips of western red cedar; the smooth silver bark of the cascara; and the shingle-like discs running down the trunks of Sitka spruce. 'Sound' leads us to stellar jays, crows and a downy woodpecker, and for 'smell', I recount the citrusy scented Douglas leaves, and Gary points out the sweet-smelling sap exuding from its bark, 'The First Nations used it as a sealant,' he tells us. Forest bathing is a good lesson in how to 'read' a forest too, and the distilled result a recognition of its multifarious components; I'm not sure I would have picked up on so wide a variety of nature had I not split my attention in such a deliberate, methodical manner. The Murdocks no doubt gave me a fairly horticulturally embellished version of authentic Japanese *shinrin-yoku*, however the meditative approach at its centre reflected what I understand of the practice.

By the end of the walk there remains one sense still to be engaged. During the morning jaunt I'd noticed Gary foraging little fragments of the forest and collecting them into a jar. Finishing the amble back where it began near the forest entrance, we sit down on a dusty log to conclude discussions. Gary brings over a steaming flask and pours its contents into mugs; now we are to *taste* the forest. We all drink. The flavour is earthy yet fruity, like a light berry tea when the bag has been extracted slightly early. 'I've been thinking about Cathedral Grove,' says Ronda, mug in hand. 'About what makes it so interesting. It's the plant community there – a mid-point between two contrasting plant communities. These ecoregions don't just go from one to the other; there's always a cross-over.' It's true that Qualicum Beach Heritage Forest has felt different to Cathedral Grove, and the plants here are not entirely similar to those I encountered yesterday. Ronda has a point, and it's an important observation; there are unique plant communities that exist even in the regions between regions, localized habitats whose subtleties are still being interpreted. Awareness of the Island's fragile old-growth Douglas-fir stands is now on the increase – protecting giant trees is something people can get behind – but trees are only a small bone in the body of a forest.

Oak

Quercus contains over 450 species, making it one of the largest of the broadleaf genera. Among them are a great many deciduous species found growing in a wide distribution of forests, but there are also evergreen examples, such as the Mediterranean native holm oak (*Q. ilex*) and the Japanese evergreen oak (*Q. acuta*). *Quercus* is a member of Fagaceae, together with the beeches and sweet chestnut, whose nuts typify the fruits of the forest. Aside from the great wealth of folklore, cultural history and environmental importance surrounding oak trees, they have an enormous variety of useful applications. Their durable wood has long served as an ornamental and construction material, remaining a popular choice for timber-framed houses today. Acorns are used as food for pigs and wild boar, and the spongy bark of one particular species, *Q. suber*, is the source of the dependable material, cork. As a host tree, both sessile (*Q. petraea*) and English (or pedunculate – *Q. robur*) oaks support a vast wealth of insects and fungi, making them a vital resource for wildlife.

English oak (*Quercus robur*)

Black Truffles and the Mediterranean Oak

In late 2017 the British media made an exciting announcement: the UK had just produced its very first black truffle, grown in the soggy soil of Monmouthshire in Wales. It seems that all our global efforts to warm the planet – with our intensive farming, coal fires and competitively-priced air travel – has at last paid off, and this hallowed, highly prized mushroom-of-the-deep now deems the Welsh climate to be satisfactorily cosy. Monmouthshire's truffle was just short of a decade in the making, a combined achievement of Stirling and Cambridge universities. Collaborating with genetic experts, MSL (Mycorrhizal Systems Ltd), researchers inoculated holm oak saplings with the spores of black truffle, planted them out and waited to see what would happen. Nine years later a black truffle was born, pungent and round; dug out excitedly by a trained sniffer spaniel called Bella. The event made national news and was celebrated in equal measure by the horticultural and broadsheet press, though I have to say I'm a little disappointed that none used the headline, 'Truffle Breaks New Ground'. This was good news for Britain, not least because black truffles are a valuable commodity – one of the most expensive ingredients in the world (in certain markets they can fetch up to £1,700 per kilo). But for the Mediterranean, where truffle yields have been falling dramatically due to increasingly dry summers, this news only added to an already growing concern. In oak forests either side of the Pyrenees – once a stronghold for this wild fungus – black truffles are becoming scarcer each year, and the fact that they now grow as far north as the UK is pretty damning evidence for climate change.

On the other hand could such a development merely imply that black truffles will tolerate a wider range of climates than first assumed? This is the question currently being investigated by Marcos Morcillo and his fellow researchers in Catalonia. Concerned with all matters mycological, their biotechnology company, Micofora, is based in the laboratories of a grand old agricultural study centre outside Barcelona. The centre is now run by IRTA (Institute of Food and Technology Research), part of the government's Agricultural and Food Department, but for many years Micofora have employed its glasshouses in the study of truffle mycology. As one of their lead researchers, Marcos looks at everything from truffle DNA to soil culture; he regularly consults on edible mushroom cultivation and hosts mycology conferences attended by biologists from all around the world. Most crucially however, with Spain's wild truffle yields falling steadily in recent years, Micofora's attention has turned towards truffles as a commercially farmed product, and to developing the new technologies that will maximize future crops.

But mushrooms aren't like seeds; you cannot sow a truffle. These little aromatic balls are an authentically sylvan product, dependent on the roots of trees just as they are on the soil around them. There is a symbiotic interplay between fungus and tree whereby each one feeds the other. Fungal roots (mycelium) gather nutrients such as nitrogen and phosphorous and exchange them for sugars produced by the tree through photosynthesis. For the black truffle, holm oak (*Quercus ilex*) is a typical companion tree, a species that dominates much of the Mediterranean landscape. But better still is an inland subspecies of oak native to drier European uplands, *Quercus rotundifolia*. It's a drought-tolerant tree with an attractive, jagged-edged leaf and, more conveniently, thrives in a similar growing environment to the black truffle. Micofora grow a great deal of these trees for the purposes of research, inoculating them with black truffle spores and bringing them on under glass, but they also sell commercially to farmers wishing to establish new truffle orchards; something now very much on the increase. Understanding a little already about the bond between truffles and oak trees, yet knowing next to nothing of this kind of fungal agriculture, I was interested to see how Catalonia's truffle supply was shifting from a foraged to a cultivated medium. I contacted Marcos and arranged to visit IRTA, asking also if he knew of anyone who could take me for a truffle hunt: if there were any oak trees in the region still concealing wild truffles at their roots, then I wanted to have a go at finding them.

I arrive at IRTA early in the morning to meet Marcos. It's January but the sun is warm and bright, alighting on the grand façade of the research centre's former agricultural estate buildings. They stand elegantly above IRTA's many glasshouses and test fields, like a manor house over a walled garden, not the shabbiest of plant nurseries by any means. Marcos greets me warmly by the staff entrance and leads us into the facility. I'm shown into a glasshouse containing work benches crammed with young oak trees. There are 70,000 saplings in total, 5,000 to a bench, all grown from *Q. rotundifolia* acorns around

this time last year. With their compact, spiky little leaves the saplings could easily be mistaken for holly cuttings, although their colouration is lighter and a shade of dusty blue-green. The leaf spikes of holm oak (*Q. ilex* – 'ilex' is the botanical name for holly, resulting in the tree often being referred to as 'holly oak') are accentuated in *Q. rotundifolia*, and this is one of the ways it can be identified as a subspecies. Last year, at a few months old, each one of these little trees was inoculated with black truffle spores, and now their roots have been colonized by the fungus. Marcos lifts out one of the saplings and opens its container, revealing a healthy mass of thin, white roots spread out inside. He demonstrates how it is possible to identify the coating of truffle mycelium on the root tips, pointing out those that are thicker and browner in appearance. Under a microscope, he tells me, every combination of tree and mushroom has a different structural character, meaning that they can be distinguished from one another even at this early stage.

Marcos believes that the scarcity of wild truffles these days relates directly to the decreased amount of traditional woodland practices in the region. Coppice workers and charcoal makers had long been present in the evergreen oak woods of Catalonia, and large areas of woodland were managed for timber. As this kind of industry was gradually replaced with modern fuels and forestry during the 20th century, management in the woods declined and much of the canopy subsequently closed up, blocking sunlight from the truffles. (Black truffles in particular require lots of direct sunlight and will therefore suffer under a dense canopy of trees.) There were other knock-on effects too, such as the increase of wild boar, for whom truffles are a favourite food. Woodland management used to keep populations at bay, whereas now they are far freer to roam. Marcos tells me that in Catalonia it is quite common for a truffle hunter to have a sideline in boar, obtaining a licence to shoot them for sport during the season. And because boar are so prevalent in Spain, their hunting season is one of the longest of any wild animal in the country.

Of all these concerns however, it is climate change that Marcos regards as posing the greatest threat to the Mediterranean's wild truffles. 'The future is farming,' he says with frank resignation. 'This year has been the worst year in decades and the driest winter for almost a century. Black truffles need at least a couple of thunderstorms each month during the growing season, yet recently there have been so few. The black truffle really is disappearing in the wild, and farming has to take its place.' Sure enough, Micofora's 'mycorrhized' young oak trees will soon find new homes in Spain and abroad, on farms where an investment is being made for the future. Despite such grim prospects for Catalonia's wild truffles however, there are still people out foraging for them. Marcos put me in touch with Albert Boixader Faja, a truffle hunter based further north in the Catalonian district of Osona. Between November and March each year Albert and his two trained sniffer dogs – a pair of border collies – spend time searching for black truffles in the hilly terrain south of the Pyrenees mountains. Albert has kindly agreed to have me join a morning's truffle hunt up there, arranging to meet me in Vic – a town about an hour away and a central hub for the Osona region. 'See what you can find,' jokes Marcos when we say goodbye, 'It's a beautiful place to walk at least, if nothing else!'

The improbability of my hunt 'coming up truffle' only makes me more determined to unearth them, and when I meet Albert in the centre of Vic early the next day, I'm already feeling pretty pumped. If only all mornings had the promise of this one: a wild fungi forage – in the oak forests of Catalonia – with a pair of truffle-hunting collies. We drive out of town and head north-east into the countryside. Winter sunlight picks out the foothills of the Pyrenees to the west, and a few snowy peaks are intermittently highlighted in the distance. Our journey continues for just under an hour, climbing gradually into higher ground and among the dark trunks of holm oak. The surface of their evergreen canopy undulates smoothly in places, almost like outcrops of grey-green moss; it's a beautiful early Mediterranean morning. Albert has brought a black truffle with him to show me, found a few days ago in the same area we're headed to. He pulls a small ball wrapped in tissue from the glovebox and immediately its aroma fills the car. The smell is amazingly strong, full of damp earth and wild garlic. I wouldn't class myself as a truffle lover, but I can appreciate a fragrance so distinct and hefty as this. As the road narrows we curve up onto the side of a hill and pull into a clearing. We get out, open the boot and the dogs spring to the ground. Albert introduces them as Guinness and Trufa. 'Busca, busca!' he shouts, 'Go search!' and the pair sprint eagerly off over the grass.

This clearing is a good spot to search, Albert explains, as it catches sunlight that falls in between the oaks, receiving a good level of warmth. The elevation is also very suitable for black truffles at roughly 1,000ft (300m) above sea level. But there are a lot of boar up here too he adds, pointing out tracks where they have visibly upturned and rooted through the soil. We walk over the low grass, studded with frosted verbascum leaves, and weave around the trees. Within minutes though the dogs are side by side and digging furiously at the ground a few feet away. We run over to the spot and Albert pulls them back, inspecting the hole. Kneeling down he pulls up a handful of earth and removes what looks like a small stone, which he dusts off and hands to me; a little black truffle, just under an inch or so thick. Putting it to my nose the smell is extraordinarily pungent, stronger even than the one in the car. There's no mistaking this unassuming little object for anything other than a truffle, though without that overpowering scent it's a needle in a haystack, a lump of earth in an earthy hillside. Albert rewards the dogs with a treat and they're off again with excited, swinging tails.

Our walk takes us in and out of the oaks and over the rise of the clearing. Both the common holm and *Q. rotundifolia* are present in a great number here, and a few smaller *Q. coccifera* too. Among them we find little clumps of box and berry-studded juniper with their armour of needle foliage. At one stage the trees thicken in a dark cluster before suddenly giving way to a dramatic panorama. The view, seen from the cliffs of Morro de l'Abella, overlooks an enormous reservoir which, as Albert points out, currently echoes the dry winter in its low surface level. When the dam was first built here in the early 1960s a small town was purposely flooded, and during dry spells like this, a little further round the lake, the remaining spire of the Church of Sant Romà can be seen protruding from the water. Back in the forest however, Guinness locates yet more small truffles, which we excavate with anticipation, including an example that has become rotten inside. Albert breaks this one in half to reveal a petrol-like odour; a foul, musty smell, similar to fermenting fruit. Truffles like these would taste as they smell, he tells me, discarding it and wiping his hands. In the local markets truffles are sold by the kilo but there is also a grading system in place. The spherical, golf-ball shaped truffles fetch the highest price; the sort that can be grated evenly across the centre by Europe's gourmet chefs in restaurants that can afford them. One of the objects of Micofora's truffle research is to understand how this rounded shape can be cultivated on farms with a greater success rate.

When my time with Albert and the dogs draws to a close we've managed to find four little truffles. It might not be a bounty of award-winning, spherical beauties, but the experience has been a lot of fun. Across the morning, as tends to be my habit, I've squirrelled away a pocket full of acorns. A few still hang from the trees at this time of year and some of those that have eluded snaffling by the boar remain scattered on the ground. I roll one in my hand, pressing my thumb down on its tight, robust shell. It's hard not to admire an acorn for the enormity of its achievements: not only is it the foundation stone of the Mediterranean's forests, but of a whole subterranean world as well; a mysterious world bearing valuable truths, more valuable than even the largest of truffles. I'm reminded of an essay by the nature writer Richard Jefferies, written not long before his death at the end of the 19th century. Standing on the boundary of a ploughed wheat field in winter, Jefferies considered the great wealth and legacy wrapped up in a handful of little wheat grains. 'Wherever they are there is empire,' he wrote, proclaiming wheat to be the ultimate currency of civilization. 'They are not very heavy as they lie in the palm, yet these little grains are a ponderous weight that rules man's world.'[1] At its heart, the essay is a lament for a diminishing way of life – that of England's rural farming community in which Jefferies was brought up. At the turn of the century it was the ever-increasing march of modernity that threatened his wheat fields and beloved natural world, but today we face the effects of another century's industrial advancement, and the future seems equally uncertain. It may not reveal yet in the green oaks of Catalonia's hillsides but there is change below the surface. As the climate warms and the soil dries, a crucial bond is quietly slipping away.

The undulating path leading to Wistman's Wood is flanked by the charred remains of gorse brush. Their twisted stems, cut short and blackened, are a clear indication of a 'managed' landscape. These are the remnants of a recent 'swale', a process whereby overgrown and dead material (primarily gorse) is systematically burned, provoking the regeneration of grassland. Swaling on Dartmoor National Park in South Devon has been practised for centuries, carried out by its respective landowners under controlled conditions. In recent years, however, its suitability for protected landscapes has become a subject of debate, an effective farming practice permitted under terms relating to conservation. While the annual clearing prevents wildfires and improves habitat for ground-nesting skylarks, it also maintains an open, grazeable grassland for livestock; an exercise therefore of compound motive. Burned gorse, if you like, sets the tone for Dartmoor as a whole: a wild and dramatic terrain, offset by its agricultural heritage.

It is one of the wilder spots within Devon's vast and rocky moorland that I've come to see: an isolated wood at the very centre of Dartmoor. Wistman's (or 'Wise Man's') Wood is 9 acres (3.6ha) of strikingly gnarled and stunted pedunculate oak trees, known best for its resemblance to a truly enchanted woodland. As a rare example of native upland oakwood, this small and ungainly grove of trees is one of Dartmoor's hidden gems, set above the West Dart River and shielded in the gully between two unassuming hills. The wood's solitary position within its barren surroundings is the first of three factors leading to its bestowed notoriety. The second is the eerie, low-lying and moss-clad nature of the oaks themselves, and the third its association with pagan tradition and druid folklore. Aside from the many recommendations I've received from fellow nature enthusiasts to visit this unique and attractive forest, almost all accounts of Wistman's Wood come laden with conjectural haunting. There are tales of ghostly processions passing through the trees at night, and it is said to be the home of the 'Devil's hounds', mystical black dogs who roam the wood, thirsty for human blood. For me, the combination of horticultural niche and paranormal pomp formed too irresistible a lure, so I arranged a trip for early spring and came to experience the oaks for myself.

Dartmoor's Pygmy Forest

The walk from road to wood stretches just over a mile, the course of which reveals further signs of both wild and managed Dartmoor. Tracts of pasture are broken into large paddocks by dry-stone walls, rising naturally with the climbing tor. Foxgloves protrude from tussocky grass and a hidden wren calls nearby. As the stone path summits at the crest of a hill a view opens up and the path ahead becomes more obvious. Observed from this elevation, one can trace the line of the West Dart along its shallow path until the river disappears behind trees, and the adjacent track stops in front of them. Wistman's Wood ahead looms compact and dense. Being late March, leaves are only opening on the earliest of deciduous trees; the distant oaks are still some weeks from breaking their winter dormancy. The absence of fresh leaf gives their branches a russet, wintry tone, and the trees form a discernible clump; together a tangible, contained little forest.

At the entrance to the wood I'm met by a small group of Dartmoor ponies. Like the stone walls and skylarks, they are synonymous with this landscape, permitted to wander freely by their respective owners. The ponies are affable, tough little things, and their presence on the moors has been documented for hundreds of years, first as workers, now as free-range grazers. They shift a little as I pass, and one particular individual, a tubby bay, confidently approaches, hopeful for a snack. But at the threshold of the wood is where the ponies remain, as they are prohibited entry by another of Dartmoor's intrinsic components, a carpet of uneven granite boulders. Nearly all of Dartmoor sits on a granite substrate, and it is thought that the historic mining of this rock is one of the reasons for the ponies being brought onto the land. Wistman's Wood itself thrusts from a clutter of granite boulders, its oak roots weaving down beneath and through the crevices between them. The stones therefore, uneven and difficult to negotiate, form a protective defence against grazing mammals, allowing the trees to flourish within, inaccessible to hungry livestock. But in turn, the wood is no less of a challenge for inquisitive humans, something that becomes clear as I approach the palpable barrier dividing grass and stone.

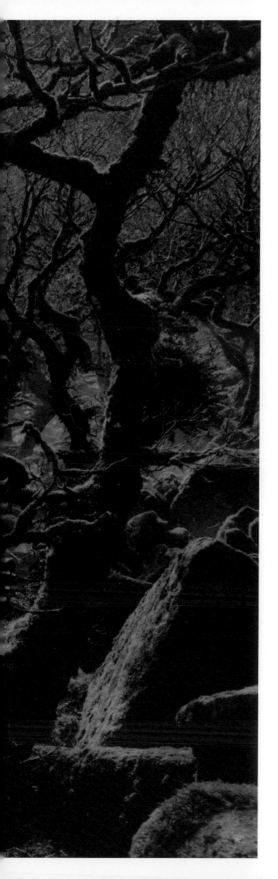

'To enter a wood is to pass into a different world,' wrote the late nature writer Roger Deakin, a sentiment never truer than of Wistman's. The combination of eerie oak and cumulus boulder is a striking visual contrast to the world outside it. A chaffinch, present and loud on a bough of a circumferential tree, stays with the ponies as I make my way in. Inside, the wood is quiet. As the granite obstructs animals' passage, so do the oaks temper the wind; a few feet in and already the air is still. There is in fact no movement in the trees at all. Moss hangs undisturbed from twisted branches, interjected at points by motionless fronds of epiphytic ferns. There is a sense of a preserved environment, reminiscent of an Edwardian glass display case, although missing its taxidermy centrepiece.

From below the wood the West Dart is faintly audible, emphasized by the silence in the trees. Without leaf break the canopy is more open than it appeared from the path, and intermittent gaps in the clouds above allow occasional bright illuminations. With these sporadic sunbursts the entire colouration of the wood is transformed: the dull green moss morphs into a striking yellow, while the oak trunks and branches lift to ochre-orange. The stones also light up and are thrown into relief, revealing distinct and irregular shapes. In such a light it becomes difficult to grasp any sense of a supposed sinister, underworldly atmosphere. The oaks are quite stunning in the sunshine and their irregular physique is a unique and entrancing spectacle. I try to imagine, for the sake of due deference, what it might be like to visit this wood in less genial weather. Perhaps in winter, or perhaps at dusk, without sunshine to hamper the ghouls. All the same, Wistman's Wood continues to feel remarkably 'unspooky', albeit lacking in visible fauna.

In my reading prior to visiting Dartmoor I did learn of the presence of one creature of particular abundance in Wistman's Wood. Clambering carefully over the granite I am conscious of the large population of adders known to reside within the stones. I'm grateful again to have arrived in early spring, too early in the year for their emergence. The adders are perhaps another factor leading to the wood's ominous notoriety – perhaps even its absence of vertebrates. However, what this little forest lacks in faunal interest is countered

by floral surprise. Appearing between stones I spot the dainty white flowers of wood sorrel (*Oxalis acetosella*), their angular leaves splayed close to the ground. There are numerous runs of bilberry (*Vaccinium myrtillus*), particularly in the higher reaches of the wood. These dwarf shrubs are covered in red springtime flowers; little pods that will form rich edible berries later in the year. The oaks themselves support a great wealth of botanical life inside their branches. Sprouting from one individual are arches of bramble, flowering woodrush and the trunk of an advantageous rowan tree, not to mention a substantial clothing of mosses and colourful lichen. After the oaks, rowans (or 'mountain ash' – *Sorbus aucuparia*) make up the bulk of Wistman's trees. They are few in number by comparison; however, some large and attractive examples can be found (that is until they are, like the oaks, checked by the moor's sweeping winds). Those sympathetic towards the mythology of this wood might read the rowan's presence as further confirmation of its mystical nature. Not only does the rowan tree possess its own folklore heritage, but those found growing from the crevices of larger tree species are also deemed by some to ward off the presence of witches. For the more sceptical and perhaps ecologically inclined, the rowans are a typical companion plant found in pedunculate oak woodlands, especially of this topography and altitude.

The origin of Wistman's Wood is, as one might expect, not entirely clear. A study carried out by English Nature suggests that the oak grove, although having changed over years of grazing and regeneration, may be a surviving enclave of a much larger natural oakwood[2], possibly dating as far back as 7000BC. It is likely that this parcel of forest was consciously spared as surrounding trees fell to grazing land. One might attribute its survival to the boulders beneath the wood, leaving the land unsuitable for conversion into grassland. The report certainly suggests that there has been extensive and varied grazing in and around the wood over the centuries, and that this will have impacted in some way on the oaks' development and growth. But typical of most 'pygmy' forests, the arboricultural dwarfism is more directly the result of environmental factors, such as poor soil and high altitude. Wistman's Wood grows on land owned by the Duchy of Cornwall, which encompasses some

of the highest of Dartmoor's tors. Residing at around 1,300ft (400m)
above sea level, it is one of Britain's most elevated oak forests,
and the stunted tree habit makes it a true rarity of its kind. In 1964
the forest was selected as a Site of Special Scientific Interest, now
protected by law and conserved by Natural England.

Before I leave to retrace my path back to the road, I sit for a moment on a larger protruding boulder positioned towards the outer edge of the wood. I'm reunited with the singing chaffinch and see that the bay pony is again within nosing range. The sun is going down behind a wash of cloud and it seems a shame not to watch it disappear. The trees gradually thin out near the lower reaches of the forest, with a few individuals standing alone just outside. Their form appears even more unique and misshapen when viewed in this way, as isolated individuals, removed from the group. The sun sinks lower and as I get up to leave it passes unexpectedly through a gap in the clouds. The sudden burst illuminates the entire wood and the boulder I'm sat on lights up beneath me. The granite sparkles, as it does throughout the forest, shining like glitter from a surface of scattered stones. With the sun on its final glow before setting behind the hill in front, it's noticeable for the first time that in fact Wistman's oaks appear to be facing it, leaning in towards it. Growing out from the hillside gradient they are a congregation of sorts, like worshippers in an auditorium, with arms reaching forward. I stand with the trees, drawn into the fading light, as if enacting some quiet springtime ritual. Perhaps it is the oaks themselves, rather than any ghoulish entity within, that endow this forest with its unnerving, mysterious figures.

Juniper

Junipers are a curious group of tough yet gawky trees. They are capable of growing in terrain far too hostile for most species: sheer rock faces, high altitude peaks, volcanic islands; and in bitter, sub-zero climates. The common juniper (*Juniperus communis*) certainly lives up to its name, being the most widespread conifer in the world. However, the ability of the species to adapt to unconventional environments has led to a somewhat unconventional physicality, and junipers tend not be regarded as the most ornamentally attractive conifers. Of the 50-odd *Juniperus* species, few are particularly tall; the Rocky Mountain juniper (*J. scopulorum*) can reach a little over 60ft (18m), as does the African species, *J. procera*, but the majority are medium-sized trees or low to prostrate, creeping shrubs.

Juniper foliage is notably aromatic, especially when crushed; some varieties smell sweet, while others are pungently acrid, varying also according to personal opinion. The tree's finest asset, however, is its berries: fleshy seed cones that develop from light green to deep blue and purple. Juniper berries have many uses in cooking, from spicing casseroles and soups, to being pressed for their essential oil. Flavouring gin of course is the most well-known, without which the drink would not exist.

Common juniper (*Juniperus communis*)

Islay Botanicals

A short distance into the wind-levelled grass stands one of Islay's hooded crows, unperturbed by the pair of lapwings charging it from above. Their swooped assault carries high-pitched squeaks of alarm and distress, but the crow continues in its destruction of their ground nest all the same. Positioned at the edge of the field I watch the scene play out for a moment; these are the first lapwings I have come across on the island and, despite the unsettling circumstances, they are a joy to see in flight. Their expansive, paddle-like wings beat like a butterfly's, enacting agile, swift runs at their adversary. But beyond the visual spectacle, to witness this episode is also to be reminded that such events occur often in the wilder places of Britain, irrespective of a human audience. The islands of the west coast of Scotland are home to vast and diverse colonies of some of nature's hardiest representatives, and sudden flourishes of avian activity like this are one of their defining charms; scenes one can so easily walk into or upon. I turn again to the narrow road leading north along the field. The air is cool; refreshingly so, and such a contrast from the muggy, claustrophobic London I left behind early this morning. An out-to-sea scent, collected by the breeze, is scattered over Islay's sand dunes and fields. It lures me on towards the land edge of Kilnave, where I am told I will find a medieval chapel with wild juniper springing from its ruined walls.

Carl Reavey, of Islay's Bruichladdich Distillery, has a particular interest in the island's scarce and hard-to-reach juniper trees. It was his description of their dramatic location that enticed me into making the trip in the first place. I had made contact with Carl via The Botanist, an artisan gin made at Bruichladdich, hoping for a few comments regarding juniper berries and gin making. Speaking with Carl I realized the fortuitousness of my lucky strike. The Botanist is not only a diversely botanical (and, as I later found out, markedly flavourful) gin, but it is infused with herbal ingredients foraged and harvested from Islay itself, making it a spirit of somewhat unique and exceptional provenance. Carl spoke enthusiastically of the island's ranging habitats: the wildflower-laden river courses, boggy peatlands, sloping clover meadows, all direct sources of the Botanist's infusions. But crucially, he relayed the niche and isolated corners where spiky juniper limbs crept, horizontal and thick, shielded from the elements. Common juniper (*Juniperus communis*), he explained, was once widespread on the island, until the arrival of sheep. During the Highland Clearances of the 18th and 19th centuries, farming land was replaced by extensive sheep grazing, having the effect of stripping all but the craftiest junipers from the landscape. Those left standing (if 'standing' is the right word for such a shrubby, low-growing tree) remained only in remote positions, unhindered by opportunistic herbivores.

The roofless stone chapel at Kilnave appears a little way off the road. It sits out towards the seafront with a circular wall enclosing its ruin and company of gravestones. The setting is oddly reminiscent of that described in *Ring of Bright Water* by the author Gavin Maxwell. I think of Maxwell's Camusfeàrna; his dwelling by the sea – a similarly isolated building in a similar Scottish landscape, only encircled by water instead of stone. The chapel, at a distance, shares what Maxwell called, 'that strange look that comes to dwellings after long disuse ... not produced by obvious signs of neglect.'[1] This 'strange look' may be further attributed to the chapel's semi-obscured appearance, hidden by the sloping gradient between road and sea. But perhaps it is a reflection of Kilnave's bloody past, whereby 30 men are said to have been trapped and burned alive inside the chapel, a definitive and brutal end to a 16th-century battle between opposing island clans. I follow a trodden grass track down to the chapel, referring to Carl's notes. He pinpoints the top of the remaining wall as home to the prostrate juniper. Unhitching a metal gate I enter the graveyard, and in my eagerness to reach the ruin I almost skip past Kilnave's historical stone cross, a landmark thought to predate the 13th-century chapel by a staggering 700 years. Ducking beneath a low arch I pass through two feet of solid stone wall and find myself standing in the centre of the grass floored chapel.

At first sight the juniper is so heather-like in appearance that I disregard it and continue to search the room. Impressively, all four walls remain sturdy and intact, comprising flat stone of irregular sizes. Here and there creeping fronds of *Asplenium* ferns fan from crevices like the tentacles of retreating octopuses. And studding the worn cement at the chapel's summit are dull flower heads of plantain and yellow bedstraw. I return to the heather and inch up closer so

that the foliage is just within reach. The sweet scent of juniper, when crushed in the hand, is unmistakably familiar. I know right away that this is of course not heather, not least for its sharp, needle-like leaves. Tracing the branch back to its body, and then further along the wall, I spot where it appears to be rooted, a lower-lying shelf of wall sprouting rugged, splintered stems. The plant is significantly larger than it first appeared, densely matted all over, maintaining a position low down with the stones. Testament to its robust genus, Kilnave's juniper demonstrates its ability to grow in the toughest of environments, altering its habit to combat challenging conditions. In a shielded environment the same plant – the common juniper – may be found perpendicular, rather than prostrate, displaying a figure more conventional of a tree. It is this high environmental tolerance that contributes to *J. communis* holding the title for the most widely distributed conifer in the world; it has the genetic benefit of being able to read a landscape, and morph to fit its mould.

A small gathering of oystercatchers with bright orange, elongated bills, call from down on the sand; their distinctive 'chinking' carries across the rock-strewn beach and up to the graveyard. The creeping juniper at Kilnave Chapel is one of the most interestingly sited examples of The Botanist gin's iconic ingredient – a squatter in a windswept ruin, looking out across the waves. However, it would be wrong to suggest that this lonesome plant is at the mercy of unregulated pilfering. The demand of commercial production in line with The Botanist's popularity requires a very large quantity of juniper berries, and such a demand could never be met by the island's junipers alone. The next morning, Carl takes me on a tour of Islay's botanicals, explaining that, where the juniper is concerned, only a symbolic amount from the island is used in the making of their gin. It is foraged in such a way so as to keenly preserve what little remains, with the hope of encouraging the plant to continue its gradual repatriation.

There are 31 herbal components that make up The Botanist, 22 of which (juniper included) are found growing wild on Islay. How one exacts an equation for the balancing of their quantities, let alone deciding upon the ingredients themselves, is a process beyond delineation; something I would assume to be the result of a certain talent for these things, and many years' trial and error. For gin to be gin, the predominant ingredient must, by law, be juniper berries, this is always the overriding flavour. But the options are limitless when it comes to experimentation with further flavourings, even if you're going to restrict their provenance to the shores of a single island.

Pleased with my successful orienteering of the previous afternoon, Carl is eager to show me another of Islay's inconspicuous junipers. Having only just arrived back from a trip to the Orkneys he wastes no time in taking me out on a plant hunting excursion. We leave Bruichladdich, driving a short way east along the coast before cutting straight up through the island. Carl navigates Islay with that enviable familiarity found in those whose occupation makes it a necessity. Various and interesting professions have led to an extensive knowledge of the landscape, including tour boat operator and editor of the local newspaper. Now in his marketing role at Bruichladdich Distillery and in working with The Botanist gin in particular, Carl has cause for even further preoccupation with the details of this remote and stunning island. Our route takes us up and over Islay's cluster of hills and when we reach the other side we pull over by a course of green and lush growth running parallel with the road. It is clear from the straight line of plants, and by their contrast with the surrounding grass, that they follow the path of a freshwater stream flowing down towards the sea. Carl points out the budding stems of meadowsweet (*Filipendula ulmaria*), rising above a profusion of foliage. He walks a short way in, welly-clad, and searches the damp ground. Reaching down he pinches a low-growing herb and brings it over to me. The opposite, oval leaves are those of water mint; wonderfully fragrant and fresh with the stream's cool running water. Meadowsweet and water mint are two of the Botanist's '22', both growing in abundance across the island's diverse topography.

As a gardener interested in wild flowers I can relate to the joy of stumbling upon unexpected floral discoveries. But there is something even more exciting in locating the particular plant you are looking for, having sought it in the environment of its origin. I imagine there is a similar thrill for birdwatchers of an equal avidity, seeking a wild and concealed species in the landscape deemed to host it. Such experiences affirm with quiet triumph our endeavours to interpret a diverse and complicated natural world. 'When he suddenly flew past me,' wrote J. A. Baker, searching endlessly for the peregrine falcon of his Norfolk countryside, 'I was lifted to joy on the surge of his wings.'[2] If afforded the luxury of time I would have taken great pleasure in the challenge of tracking down each of the Botanist's 22 wild herbs. However, given the constraints of this visit Carl, knowing all of the spots, shortcuts us through some of the key habitats. We visit a flat peatland, rich in aromatic bog myrtle, in which we find spotted orchids and sprawling sphagnum moss. In a hillside mire we pass mugwort and marsh cinquefoil, and along crumbling road verges are the pink flowers of creeping thistle. The stout outlines of hawthorn trees appear throughout our drive, protruding from Islay's drystone walls. Although not yet in berry, these

will soon contribute yet another botanical to the list of Islay infusions. Our final stop is to a little cove below drifts of native coastal flowers. Here, lady's bedstraw (*Galium verum*) proliferates in yellow swathes alongside sea pinks, white clover and purple thyme. Carl leads down a narrow slip towards the beach. I notice our approach being observed by inquisitive island residents; from the water by a bobbing grey seal, and from the sand a pair of plovers. We walk across to the opposite bank and there, growing out above us, is another horizontal juniper. This time the tree has rooted through sand, clinging to the little available soil above sheer protruding rock. It appears almost clipped as topiary, shaped like the rock itself; a bigger specimen than the one at Kilnave, yet apparently equally as shallow-rooted. Despite their woody nature (and indeed being classed as a tree), the island's prostrate junipers are very much at home with The Botanist's foraged herbs. To encounter a tree camouflaged in such a way within a landscape reminds me once again of their great diversity of form.

On the last morning of my all-too-brief visit to Islay I meet again with Carl at the distillery. Joining us today is James Donaldson, The Botanist's recently appointed, well, botanist. James heads up Bruichladdich's foraging team, having taken over from its founders, Drs Richard and Mavis Gulliver. The scientific authenticity of the spirit's 22 harvested ingredients now falls upon James to ensure, charged with the collection and preservation of these wild plants. He meets us in the distillery car park, carrying a crate of rooted juniper cuttings. Their single stems, like little sprigs of rosemary, have been potted into 3-litre containers. We drive a short way up behind Bruichladdich to a recently levelled road and unpack a mattock and spade. It is Carl and James' hope that some of these cuttings will take hold in the stone of the new verge here and help reinstate a little of the island's missing juniper. The verge is steep and absent of quality soil, a desolate setting comprised mainly of compacted stone. James sets out the pots, positioning them carefully on the ground, before excavating crude holes with the mattock. With considered planting and a little supervision, I can picture the juniper thriving in these less than inviting conditions, rising to the challenge of their familiar island landscape, and spreading slowly over the ground.

Two controversial forces run wild in the high desert outside Bend, Oregon: wildfire and the western juniper tree. They share a status as naturally occurring, native elements of the semi-arid landscape, and both display – to put it lightly – significant 'invasive tendencies'. Western juniper is one of America's toughest trees; a species that has prevailed over the harsh growing conditions of the upland desert, with sprawling roots, a slow growth rate and a changeable physical habit. Historically, wildfires are one of the few natural ways in which western juniper has been kept in check, but as cattle farming and urban expansion over the years have led to a suppression of fires in the American West, the tree has steadily extended its range, and increased dramatically in number. In parts of rural Oregon the species is now something of an unstoppable force, seeding down from its natural rocky range, spreading across sagebrush and grassland pasture, and intercepting what little rainfall would otherwise enter the soil. It isn't often that one hears a story of natural forestation; of woods *appearing* rather than disappearing, yet juniper is doing just that, with serious implications for the ecosystems upon which it is encroaching. Controlling such advances poses an interesting challenge: not only does an enormous root system make western juniper tricky to remove in practical terms, but the tree also has a rare quality of being both native and invasive; it is protected in some areas as an ancient indigenous flora and loathed in others as an intrusive pest. And here is where the two controversies intersect: fire is key in suppressing the spread of invasive juniper, but fire is an unwelcome guest on the doorstep of growing towns, cities and farmland. In the western states, human interplay is caught up in an ecological whirlpool, whereby holding one natural force at bay has affected the advancement of another.

Juniper in the American West

When I arrive in Bend, a sullied haze reddens the sun; it sits like a plastic orange filter, masking-taped over the sky. Fires have been burning out in the surrounding juniper brush, 'prescribed burns' controlled by teams of firefighters and professionals from the local Forest Service. These small, manageable fires are a systematic way of reducing the build-up of woody material that, should a wildfire ignite, would burn fiercely and potentially out of control. Vigorous juniper is the main culprit for the woody deposits, a tree that surrounds Bend on all sides like the saguaro cacti of Tucson, Arizona. They range from young, conical-shaped little shrubs, to gnarled and disjointed mature trees, some of which are very old indeed. Coming down from the plateau of northern Oregon I watched their number steadily increase: they began in little colonies, dotted on either side of the straight road, before gradually filling out over hillsides and condensing towards the edge of the Great Basin at Bend. In town it takes me a moment to realize that the scent I'm smelling – a slightly sweet yet damp must – is the juniper itself, wafting lightly in beneath the haze and perfuming the city. Even my hotel room hints at the aroma. The tree flourishes in the high desert landscape here, a terrain that isn't technically desert but closer in nature to grassland steppe. It's arid, but not too arid, and results in part from the leeward rain shadow caused by the Cascade Mountains to the west. High desert exhibits a wonderfully simple colouration: a palette formed of *Artemisia* silver, rabbitbrush yellow and the dark green and brown of western juniper. Interweaving the shrubs and scattered trees are low grasses and sedge, such as Indian ricegrass (*Achnatherum hymenoides*) and threadleaf (*Carex filifolia*); all green in spring but now a blazing summer gold. Eager to see some of this wild terrain and to get in amongst it, I head out under the glowing evening sunlight to Smith Rock State Park.

Smith Rock lies 25 miles (just over 40km) north of Bend, formed from an outcrop of volcanic and basalt rock. For climbers it is perhaps Oregon's number one spot, offering some of the toughest and most scenic climbs in the state. In terms of landscape however, Smith Rock is the embodiment of the Wild West; you couldn't ask for a more iconically dramatic setting. An initial view looks down over a dry and commanding valley with jagged and monolithic stone rising above the winding loops of the Crooked River. The river banks bleed green where water has irrigated the dusty earth, a lush inconsistency within the local vegetation. Once down there, a closer look reveals swathed sprays of wild rye grass and tansy mustard interspersed with clumps of erect *Equisetum*. Astonishingly, as I follow the river, a beaver swims out across the water beside me, branch in

mouth. It glides smoothly towards a small island of rock and disappears behind, ploughing long arrows that ripple downstream. Blackjack pine (*Pinus ponderosa*) borders the river here too, lofty and straight, with a skin of decorative red scales. Juniper and ponderosa pine share a common ground in areas like this, where the watershed offers sufficient moisture for the pine to prosper. Away from the water's edge however, the steppe flora quickly resumes, accented by sporadic junipers and large mounds of sagebrush. In places, fresh young sage sprigs mirror the basalt rock behind and blend silver-grey with silver-green in a mesmerizing harmony. I wade in among them, brushing out the thick, herby scent; a warm perfume so much a feature of this landscape. Last to attract my attention, wild asters appear in purple-blue and yellow at the foot of the sage, obscured at a distance by a thin coat of dust.

Smith Rock is a peaceful place – time slips by quickly and before I know it sunset has begun to darken the rock faces into enormous silhouettes, and the air has steadily cooled into the direction of the night. It's quite a way back to the car so I circle back and pick up the pace. I was forewarned that rattlesnakes live among the rocks in the valley, and that dusk is when they tend to be most active. Alarmingly, snakes do begin to appear on the path in the diminishing light, and I'm embarrassed to say that on more than one occasion – when an outline suddenly materializes in front of me in the dark – I find myself jumping to one side or skipping over them in abrupt, awkward leaps.

The next morning I'm up before the heat. I've arranged to meet Gena Goodman-Campbell from the Oregon Natural Desert Association (ONDA) for a morning's hike into the high desert. She plans to introduce me to some of the region's old-growth juniper in the nearby ecological research area of Horse Ridge, and has suggested we make an early start before the sun gets too high. As Public Lands Coordinator for the ONDA, Gena spends much of her time in the wild terrain outside Bend, campaigning to preserve its unique and vulnerable ecology. In the late 1980s, fears over the growing impact of cattle grazing and general human disturbance in these areas spearheaded the ONDA's inception, a non-profit organization bolstered by volunteer action, dedicated to protecting Central Oregon's high

desert. Gena advised bringing plenty of water for the walk, irrespective of the early start, as we would aim to locate a particular western juniper thought to be 1,600 years old, growing in a spot fairly high up on the ridge. Her dog Penny joins us for the morning, a black and white mongrel full of energy who came to Gena as a rescue, found wandering alone and collarless in the nearby prairie. As we begin our hike into the sagebrush Penny tramps up in front, lifting little clouds of light dust underfoot.

The hike begins in a plain of flat, scrubby prairie with a scattering of juniper. I find myself walking much of this stretch with my eyes on the ground; brushland may appear barren at first but the asters at Smith Rock have already taught me to expect concealed gems. Sure enough wiry stems of pink-petalled *Erodium* appear between the rabbitgrass and sage, and dry, orange tufts of buckwheat emerge from broken stone, reminding of the sea-pinks that cluster along Britain's floral coastline. Very soon the ground begins to rise and we start up towards the ridge; by now the juniper is more or less everywhere, congregating in the higher levels of the steep. By 10am the sun is blazing down and forcing sweat into our backpacks: we stop a couple of times to rest, drink water and check the coordinates Gena has been given for locating the ancient tree. By the third pause poor Penny is panting like a spluttering tractor engine.

I notice that the trees have now become a marvellous assemblage of strange and splintered shapes. Their foliage is unbalanced and patchy; distributed over limbs and branches in sporadic flourishes, while the branches themselves – a rigid tangle – splay out into shards of grooved, hardened wood. Some limbs seem to have died back completely and have been set-upon by an acid green lichen; 'Wolf lichen,' Gena points out, a good sign of old-growth juniper. 'Lichen doesn't start accumulating until the trees are a few hundred years old,' she says, turning enthusiastically to another juniper behind us. 'This is a cool one – you can see it's pretty much dead except for one branch. It shows all the characteristics of an old-growth: the bark is flaky and the top is domed, and its boughs are all twisted and furrowed.'

We get a good workout searching for the 1,600-year-old juniper, clambering up the hillside over stone and soft earth, and when we do find it – an unassuming specimen no bigger than any other we've so far encountered – there is little out of the ordinary to suggest its supreme age. The tree sits on a pile of raised rocks, a circumstance which might have contributed to its long life – western juniper, as discussed, is not immune to fire, and it prospers on ground out of reach of the burning brush. This is why its natural range is in mountainous, rocky terrain; only since fire has been suppressed has it begun to dominate grassland habitats. A 3-inch borehole in the juniper's trunk indicates where biologists took a core sample back in the 1990s. This was part of an investigation carried out by the Bureau of Land Management to determine what kind of a forest was growing up here on the slopes of Horse Ridge. Due to the splintering habit it is remarkably hard to age mature juniper, but thanks to research of this kind, more and more of Oregon's precious and irreplaceable old growth is being determined, protected and secured for the future. Wilderness designation makes it illegal for anyone to cut down trees, to do so carries a fine of up to $100,000. Previously, residents in the county had been removing ancient juniper for firewood and ornamental use, perhaps many not realizing just how rare and irreplaceable a material it was.

Resting for a moment at the base of the tree I look out over a landscape of dry grass and patchy green; an army of virile juniper stretches way off into the distance, merging with the horizon in a wash of olive yellow. Out there is the Oregon Badlands Wilderness; over 27,000 acres (11,000ha) of mostly flat, hardened volcanic terrain. The Badlands was only appointed as a wilderness area in 2009 following much campaigning from the ONDA. The week before my trip to Bend an editorial piece in a regional newspaper ran the headline: 'Wilderness Doesn't Mix with Civilization', arguing that proposals to designate yet further wilderness areas close to communities posed a significant threat, especially from wildfire. 'New wilderness,' remarks Gena, as we make our way back down Horse Ridge, 'I got a kick out of that. These areas have been wild for millions of years; it's the community that's new.' In many ways it's hard to imagine rural cattle ranches as having an overly detrimental effect on wild countryside, indeed they seem almost the antithesis of urban city sprawl. But the writer Adam Nicolson put it well when he wrote, 'uncontrolled grazing, the beautiful Western sight of a cowman with his cattle on the range, could destroy a whole life system.'[3] Nicolson refers to the impact cattle farming has had on wolf populations in the American West, and how their suppression has led to an ecological imbalance. 'Cowboys, cows, no wolves and too many deer and elk added up to ecological disaster,' he declares, a notion equally true if you replace the word 'wolves' for 'fire'. In the case of the western juniper, agricultural expansion into wild territory has turned a native into an invasive plant. It is a plant as emblematic as it is vital to the high desert ecology, but a tree that now threatens a delicate sagebrush ecosystem in which woodland has no place.

Birch

Birch trees must be among the most familiar of all the world's broadleafs. The white, papery bark characteristic of some species is as attractive as it is distinctive, lending favourably – almost too favourably – to application in landscape design. *Betula utilis*, a Himalayan species, holds true to its name given the degree to which it has been utilized in contemporary ornamental planting. But while it is praised by garden designers and civic planners for its light, airy canopy and unobtrusive root system, birch is also something of a weed. Both the familiar Eurasian birch – silver (*B. pendula*) and downy (*B. pubescens*) – are prolific self-seeders, in a way making them the arboricultural equivalent of a dandelion. The silver birch in particular, with its high tolerance for dry, poor-quality soil, is a pioneer plant, quick to establish in areas that have been scorched by fire or manually cleared.

Birch forests are widespread throughout northern Europe and often occur in areas where other trees struggle to take hold. They can be found in the taiga of Norway and Alaska, yet grow just as freely on the volcanic slopes of Mount Etna in Sicily. The trees also support numerous insect species and their seed is a staple of some of the less prevalent British birds, such as siskins and redpolls. A distinctive fungus, the birch polypore (*Fomitopsis betulina*), can often be found protruding horizontally from the trunks of silver birch, similar in appearance to a shelving bracket.

Silver birch (*Betula pendula*)

O n my desk sit three utensils made from rough-cut birch wood: one knife and two spoons. They're angular yet smooth, lightly coloured, intersected by dark, curving streaks. Each is rigid and tough, much more so than I might have expected for birch, and carved for function with purposeful cuts. Their shaping is traceable through marks left exposed, spared from any levelling with sandpaper, and like all raw wooden objects these items have an inviting tactility. I've found myself picking them up from time to time and rolling them in my palm, often unconscious of the fact while absorbed in something else. Their smoothness is oddly calming and feels good in the hand. These pieces were a gift from Sophie Sellu, a woodworker from London whose studio is not far from me in Deptford. What attracts me to Sophie's work has as much to do with the process as it does its ornate results. Despite her years of experience, Sophie carves with the excitement of a newcomer to the craft, composing each piece with a devoted absorption. Even when demonstrating her method to interested parties, she has a propensity to get carried away, lost in the task until it is completed. Witnessing an object take form through a series of fervent knife strokes is mesmerizing to watch, but even more so on account of its equally focussed sculptor.

This treatment of composition – of each piece as an individual creation – is a characteristic that has at times jarred with commercial elements of Sophie's craft. While the personality in her carvings has secured significant patronage (such as high-end department store Liberty), the conveyor-belt demands of delivering uniform commissions can temper her fluid approach. Sophie works in tune with her material, trading forceful strokes with the will of a particular wood. She might start on a project with a finished form in mind, but the resulting creation is a playoff between artist and artefact. Unlike most wood-carvers, Sophie's source material is already artefact, a pre-sculpted medium, given purpose for a second time. From

Reclaimed Grain

cabinet furniture to old kitchen worktops, most of the wood entering her studio is salvaged material, wood with a function no longer required. Carving with seasoned wood is much trickier than with fresh: it can splinter more easily and prove stubborn to mould. But the upshot of this lignified medium is its availability: green wood is not so easy to come by when you're based in central London.

It is an enviable erudition to tell apart one seasoned timber from other. Antique furniture sellers are often proud possessors of such expertise, confidently proclaiming – with varying reliability – the parent tree of a given chair, table or ornament. It is an odd thing, I suppose, that one might identify a living tree with decisive familiarity, yet struggle to recognize it in everyday objects, such as a pepper grinder or companionable bedside table. However, there are many whose primary experience of wood is in its timber form. For Sophie, having both grown up and remained, for the most part, living in the capital, this is very much the case. There are things she knows about trees to which, as a gardener, I wouldn't necessarily relate. I wouldn't, for example, know the difference in spalting pattern (colouration on wood caused by fungi) between beech or sycamore. Nor would I recognize the relative densities and heartwood colour of walnut, maple or western cedar.

The vast collection of timber is what immediately captures my attention as I enter Sophie's high-rise studio. We've arranged to meet in the final days of her residency in a Deptford industrial block, although when I arrive the packing up is really yet to begin. 'As you can see,' she laughs, 'I haven't got very far.' The little room is filled with all kinds of wood. Dark planks stand stacked against walls laden with hand-carved shapes, while others are bundled together, or balanced on furniture, bookcases and workbenches. It's the right kind of clutter: a cave of attractive odds and ends, offcuts and outcasts collected over years of salvage and donation. It doesn't take long before I am magpie-ing my way through her miscellany, lifting out blocks, lured by swirling grains. There are a few familiar pieces among the hoard; a hunk of yew, for example, glowing meat-red at its centre. But the majority are predominantly strangers, certainly unrecognizable at first glance. Stoking my enthusiasm, Sophie eases an enormous wooden slab out from behind a heap of stacked planks. 'It took me a while to notice – check out the rings.' The wood is dark but the growth rings are easily discerned. It takes me a moment too to realize that the grain rings running from side to side through the end of the sawn timber display an alarmingly shallow curve. The tree this was cut from must have been enormous, the slab itself displaying just a fraction of its colossal girth. Sophie indicates that the wood is iroko (*Milicia excelsa*), a tree that can reach over 160ft (48m) in height. She hasn't a purpose in mind for it yet and, like so many pieces in her stockpile, the wood will sit patiently until the right employment arises.

There is something intrinsically pleasing in the collecting of natural objects. The curiosities of nature have an irresistible lure for some people, who seek to surround themselves with them. I remember meeting the illustrator Peter Cross and later being introduced to the snail shells and pine cones whose intricate details had featured in familiar drawings of his from previous decades. Peter had accumulated these items on walks around his home in the southern English countryside, and stored them for closer artistic observation. Similarly, although for a less studious motive, it is rare that my mother will leave a beach without pocketing a shell or two, all of which have found a home on the various shelves, chests and bookcases of our homes over the years. I suppose it is a very human way of connecting with a landscape. Walking around Sophie's studio reminded me of my time as a head gardener in South London, and my shed tucked beneath a sycamore tree, bulging with Nature's ornaments. A journal entry from this period reads:

SEPTEMBER, 2012

Looking around the shed this afternoon I noticed almost for the first time just how many objects of the natural world I've brought in and stashed for one reason or another over the last couple of years. A small alder branch, still clutching catkin and hard-cone fruit sits on the bookshelf by my radio, and a collection of this year's giant *Allium* 'Christophii' seedheads stands in the corner of the room next to the log baskets. There are the tall collected stems of *Veronicastrum* (once purple, now brown), and the silver-powdered branches of *Perovskia* 'Blue Spire'. On my desk there are dried *Verbena*, *Lunaria* and sea-holly stems, and next to these a great jumble of items, from granite stone to jay feathers.

Working explicitly with a reclaimed medium has its obvious plus points; no trees are harmed in the making of Sophie's craft – not in a direct way, in any case. She tells me that people often get in touch and offer pieces of attractive wood; they ask if she'll fashion something from it, and then donate the rest. On one occasion Sophie was brought an entire suitcase of hardened walnut as a gift by an American journalist. Walnut is a favourite to work with, she remarks: people would buy it for expensive kitchen worktops and cabinets, but you rarely see it used for utensils and cookware. The rich, deep brown of her walnut spoons stand out on a shelf, displaying a striking opulence for such a commonplace object.

It is a wood at the other end of the colour spectrum that is being carved in the studio today. A long, light birch branch lies on Sophie's desk, seasoned and dry, in the hand weighing next to nothing. Elsewhere in the workshop larger sections of birch have been sawn and split, cleft down the centre revealing a dull and pale heartwood. Sophie sits at her desk, pen in hand, and draws the outline of a spoon on the rough surface of a birch block. Although seasoned, the wood is pure and unmarked. These cut sections were donated many years ago, offered at the time by a friendly tree surgeon as practice material for Sophie's developing craft. Until now the small logs had been stored with the rest of her offcut wood, waiting their turn to be put towards specific use and purpose. Sophie's gift from the tree surgeon chimes with a similar experience of John Stewart Collis's, a recollection he uses to sing the praises of birch wood: 'Once, having been given four freshly cut logs of silver birch, I did not burn them ... but put them on a shelf as pictures. And I assure you they held my attention for many a day.' Even those little interested in trees are often comfortably familiar with the bark of silver birch (*Betula pendula*), and at the underside of Sophie's block remains that identifiable, signature white peel. As she turns the block over in her hands it occurs to me that the wood is, in many ways, the perfect intersection between our respective relationships with the same entity; a seasoned timber retaining its natural form. Carving these pieces is a rare opportunity for Sophie to work with a 'virgin' wood; a chance to sculpt with un-sculpted material.

Taking her knife and paring down the edges, Sophie describes the wood's grain as a pleasure to work with, 'fine and subtle'; easy to carve. Very quickly a handle takes shape, and she replaces a straight blade with the curve of a traditional hook knife. The hook excavates a sump at the head of the spoon and before I can interpret her movements the shape is revealed. After some finer sculpting with short, defining strokes, Sophie inspects her work and hands me my finished spoon. 'Since I was about 14, at school,' she tells me, 'I've been making things. I was always in the woodwork shop; it's all I wanted to do.'

There are giant cedars and oaks, indeed single birch trees also, that as living, growing entities can captivate our attention. But there is something heartening in the confirmation that wood retains this same magnetism when entirely divorced from the outline of its original parent material. Like so many collected items of the natural world, there is something beyond its appearance that draws us to wood; something tactile that stands quite separate from any familiarity with its arboricultural heritage.

Pioneers of Saxon Sandstone

'Has any tree so graceful a way of throwing up its stems as the birch?' wrote Britain's foremost garden designer, Gertrude Jekyll at the end of the 19th century. 'They seem to leap and spring into the air, often leaning and curving upward from the very root, sometimes in forms that would be almost grotesque were it not for the never-failing rightness of free-swinging poise and perfect balance.'[1] I love Jekyll's admiration of birch trees; their description in *Wood and Garden* continues for another paragraph or so as she is seemingly sidetracked while walking through her Surrey copse. With a flourish of enthusiasm Jekyll describes the subtle shifts in colour – in whites – from 'silvery white' to 'milk white', and the compelling interaction between smooth and ridge-ridden skin. But this 'never-failing rightness' and perfect balance she refers to is what currently holds my attention. It is a characteristic displayed every bit as strikingly in front of me, in the birches of Saxony's sandstone mountains. All along the narrow ridge at my feet they jut from shallow rifts, reaching out over a hair-raising drop down into the wooded valley below. As if to spite the overhang their small trunks curve upwards and stand proudly erect, like a gymnast on a tightrope, as unobstructed by the precarious position as by the sudden blusters of wind charging at their branches.

The border region straddling Germany and the Czech Republic is layered with a renowned range of habitats and corresponding flora, but only birch and pine brave the ground up here, the highest of the Elbe Valley's many elevations. Now at the tail end of September the birch are beginning to turn; their leaves have become a collection of faded green and craft paper yellow. Those having sped through the season are already afloat; lifted and swirled up the mountainside from trees lower down. I brace myself against the smooth rock while passing over the ridge's summit; this is the last of three stunning climbs and certainly the most unnerving. In one or two places the floor drops away completely between outcrops of rock, leaving alarming, foot-long gaps along the path in front. I remind myself that I am not so much scaling a mountain as an embattled ruin, the remaining fragments of a cretaceous seabed. So iconic of the region, these ridges have been sculpted over millennia, gradually etched

away by the force of the Elbe river and its tributaries. Such weathering has resulted in strange and unlikely shapes on the horizon: arching bridges, huge mounds and single, top-heavy peaks, all of which rise from the dark forest like stalagmites in a cave.

Reaching the end of the ridge and with no further rock in front of me, I crouch for a moment beside a young birch, its torso only an inch or so thick. In my grasp the tree feels reassuringly sturdy, despite its youth – although age may not relate to size in this case, given how windswept the peak is. I suppose trees up here grow slowly in resistance to it, excavating a solid grip before venturing further into the sky. The smooth bark is certainly rippled into maturity in places, a feature that reminded Gertrude Jekyll of 15th-century German costumes, 'where a dark velvet is arranged to rise in crumpled folds through slashings in white satin' – rather fitting an observation for this particular location. Although almost October, it is still too warm to be wearing anything other than a T-shirt up here on the mounds. The mid-morning sun is already alive within the stone, highlighting rock formations further out in the valley and sending long shadows down through the gullies and crevices nearby. It catches the Elbe too, a little way off in the distance, reflecting brightly on the water's surface as if it were a sheet of thick ice. It's a big old river, even when viewed from this high up. One can see how it has forced its way through the forests that rise on either side, forming a deep scrape through the otherwise unbroken green landscape. The current has meandered all the way from the Polish border in the north of the Czech Republic, having risen in the Krkonoše mountains that divide the two countries. From here the Elbe will flow through Germany more or less in a diagonal line to Hamburg, at which point it will widen to an estuary before emptying at last into the North Sea. I picture this valley in the throes of winter, with the snow that will likely fall by January. True autumn is on its way though and will arrive soon enough, an event no doubt extraordinary, especially along these slender ridges, all highlighted with blazing yellow birch.

It was on a train passing between Prague and Dresden that I last encountered the Elbe's intriguing valley. This was some years ago while following the river northwards into eastern Germany. I knew next to nothing then about the river or the landscape hurtling past the window to my right, but I remember being completely absorbed in the view. A diary entry from the journey recalls snow-dusted alders and beech leaves caught in the webs of seeding clematis vine strung all along the valley. Every now and then something of the strange hills beyond revealed themselves in squinted glimpses – I'd need to come back and visit this area properly some time, and get in under the trees.

Sure enough Saxon-Bohemian Switzerland is every bit worth the return, a kind of forest walker's paradise laden with incongruous stone deposits. The name of this region is a little fantastical in itself, a compelling combination of words bringing together two territories in one. While 'Saxon' denotes the contribution made by East Germany's free state, 'Bohemian' refers to similar land situated in the region of Bohemia in the Czech Republic. Both are independently managed national parks but are comprised more or less of the same stuff, strung together by the course of the Elbe and its extraordinary natural features. The 'Switzerland' part is somewhat confusingly derived from observations made by visiting Swiss artists in the late 18th century. Writing letters home from Dresden, they likened the picturesque mountain view of Saxon-Bohemia to that of their own country, and the name subsequently stuck. But the greatly varying topography of the region as a whole is what makes it so unique. Even as one national park the total land area isn't all that vast – some 36 sq. miles (93sq. km) on the Saxony side and just shy of 31 sq. miles (80sq. km) over the border – however almost all of it is covered with forest of one sort or another. In core areas one need only walk a short distance to have passed through a number of these ecological divisions. There are steep gorges and meandering watercourses, and thick stands of tall Norway spruce. There are the high rock plateaux containing beech and fir woods; elevated yet further are the sandstone ridges running narrow like a splintered spine. What makes these environments even more interesting is that they invert the climatic distribution of a typical mountain range, turning it on its head. At the feet of the stone formations conditions are generally damp, dark and cool, whereas on top they are dry and comparatively hot. Were the valley's rock formations a true mountain range you would expect the reverse: warmer temperatures down in the low forest and cooler higher up. But the park's top elevation sandstone ridges are warm, dusty and free-draining due to the porousness of their structure, which has led them to be colonized by birch, a pioneer of tricky terrain.

Each climb up to these birch-lined summits was prefaced by a hike through elevations of the cool forest below. Staying near the small village of Hřensko I began my exploration of

the Elbe Valley on the Bohemian side, before crossing over to Saxony a few days later. My introduction to the region therefore was at the base of Pravčická brána, one of the more prominent outcrops of stone in the shape of a wide archway. Rock formations like these are often referred to as bridges, and as bridges go Pravčická brána is Europe's biggest. The pathway to the brána (Czech for 'gate') begins under dark coniferous cover. It's so dark in fact that it feels almost claustrophobic. Everything appears damp, from the slippery path to the hillside rock, which has a kind of riveted tree bark surface to it, similarly porous and almost spongy in texture. Even the air feels wet, which increases the sense of enclosure, like being shut in a cool cellar or a giant meat fridge. The path rises steeply and already I'm looking for the way out, picking up my pace a little despite the gradient. But soon enough the flora begins to change; first a maple sapling, then a beech, then patches of sunlight and a flower or two. I find splayed bells of *Campanula* with petals of deep purple, and next to them are the browned seedheads of spent campion. I continue up another elevation and here the rock really begins to show itself, sitting heavy in the leaf mould, full of nooks and narrow caves. Noctule bats are said to roost in the sandstone and hibernate here during the winter, which means that at this time of the year, slightly higher up, they'll be eyeing up the stone and preparing themselves for torpor. The air is warming now, becoming fresher. In still brighter light I reach a small grove of birch. They are as John Stewart Collis once described, 'knuckled, notched and dented', yet also elegant. Their collected white trunks recall those grouped architecturally outside the Tate Modern gallery by the River Thames in London. And now at last I'm out into the open among scattered pines and yellowed fern fronds, and the sandstone – so much drier here – matches their autumnal colouration. A view over the forest opens up between the pine branches; in the distance I can spot numerous outcrops of high, elevated stone. A short way further and I reach the brána rock itself, my head at last above the crowd of trees and able to gain a sense of geography.

Many paths connect together at Pravčická brána and in my ascent onto the flat of the summit I'm joined by other walkers. It's a popular spot and as a precautionary measure against excessive footfall the bridge itself is barred from access. Instead you walk under it, and up into the surrounding stone, which offers a great view of the brána from a peak nearby. A spiralling path winds into the ridge bringing me up through the sandstone. Close up like this I get a picture of the many kinds of weathering – fissured grooves and mottled markings for example, and holes like those of a woodpecker. There are funny little pockets too, into which a strange menagerie of plants have seeded: a sapling rowan tree, wild strawberry and a rogue Asian balsam with its yellow, viola-shaped petals.

Things feel different in Saxony, when I later cross the border; it's the same region; the same valley, yet a new set of environments. Leaving Bohemia behind I find myself entering a different kind of forest again, one equally dark and wet, but somehow more intrinsically European, for want of a better term. The character of the wood is one of moss and mushroom, it's almost swamp-like in places, with a pervading sweet smell of wood decay. There's an eerie quiet like a vacant quarry. I frequently stumble upon fly agaric fungi with their luminous red caps, could there possibly be a more iconic symbol of the enchanted wood? But there is a reason why the woods here appear as they do. The national park has a wonderful ethos that allows nature to be nature: a 'Noah's Ark' approach, whereby each plant, bird and insect is deemed equal and equally important. The Free State of Saxony has elected that the Saxon-Swiss forests here are encouraged to return, after years of arboricultural conditioning, to natural processes of decay and regeneration, ensuring that, by law, they are no longer commercially interfered with. Even what might be considered a 'disaster', such as a storm or insect invasion, is now all but welcomed; considered, rather sensibly, to be a part of nature rather than an action against it. As Richard Mabey wrote in response to Britain's Great Storm of 1987, 'catastrophes – be they disease, climatic trauma, insect predation – are entirely natural events in the lives of trees and woods. They respond, adapt, regroup.'[2] Intervention by the park's rangers is therefore also kept to a strict minimum, and for as much as possible the woods, streams and rocks are now left to their own devices. This is good news, for instance, for the many species of spider living inside the Elbe's sandstone rocks, but also for those creatures dependent on the habitat of decaying trees.

Once more, my morning is spent ascending from the valley basin. Again I pass through layers of flora varied in make-up and character by the region's vertical divisions and the microclimates they have produced. It's not long before I'm back climbing among birch and stone, though the climb is substantially lengthier than those of the previous days. Long iron ladders have been installed by the park to assist with difficult sections – particularly where there is nothing but sheer rock – and one by one they take me up onto the windy ridge. In certain parts of the climb the birch intermingle with Scots pine, but where they stand alone they are a truly lovely sight. My experience of wild birch in the past has been so defined by damp and miserable ground that it is a pleasure to see them in such a contrasting environment. Up here they are the superior tree above a valley of many, standing in command of one of the greatest views Europe has to offer.

Chestnut

This chapter is concerned with one of the most stately of broadleafs, the sweet chestnut (not to be confused with the horse chestnut) . Like oaks (*Quercus*), they belong to the beech family. *Castanea*, the genus to which the sweet chestnut belongs, contains around a dozen species worldwide but only one is native to Europe: *Castanea sativa*. The American sweet chestnut (*C. dentata*) is a particularly tall and elegant species, though over the last century it has sadly all but disappeared from its native east coast forests, due to a virulent pathogen introduced from abroad. The culinary merits of the nuts themselves need little introduction, however, both leaves and flowers contribute in equal measure to the overall charm of sweet chestnuts. The flowers appear in early to mid-summer, and comprise yellow-white spikes, like elongated willow catkins. While the sweet chestnut is very much settled in Britain, particularly in the southern regions where it has been coppiced for centuries, the tree is thought to have been introduced by the Romans. In recent years coppiced sweet chestnut poles have seen something of a resurgence in popularity, providing rustic yet functional material for fencing posts, panels and roofing shingles.

Sweet chestnut (*Castanea sativa*)

The winds sweeping up through Montseny's steep gorges are hushed the moment I enter the tree. The bitter January air they carry also abates once inside, stopped short by a foot of weathered wood; here is a shelter on the edge of a cliff, as protective as a cave and natural as stone. Besides its low doorway, which requires me to stoop to pass through, I find the interior to be complete with a level floor, window and smooth wooden walls; it is decidedly homely and, rather surprisingly, not in the least bit dank. The inner face of the trunk is worn smooth on most sides except for the section in front of me where it has been charred black by fire. The surface feels like rock, hard and cool. Standing fully upright I come nowhere near reaching the ceiling, the apex residing another few feet above my head. Up there the wood is much softer than the lignified structure around me; it is younger in age and, although no longer living, sustains a form of life in its continuing decomposition. The effects of microbial processing furnish the once living tissue with a variety of colouration, ranging from a corky orange to damp, mottled brown, and the whole surface is illuminated by a second, higher-level window further up. Around this small opening dangle strands of cobweb and dry grass, which I presume to be the remnants of temporary nests made by summer migrants. The window lower down is formed of a much larger opening, shaped like an arch and reminiscent of an imposing church window above an altar. Its smoothed edges perfectly frame a solitary conifer outside in the clearing with a backdrop of mixed forest beyond, the freshness of which carries into this wooden chamber an appealingly clean mountain air.

In the dim light I scan the rest of the tree's interior. To my left, at shoulder height, the wall curves back before continuing upwards, forming a natural protrusion one could imagine repurposed as a shelf. I find a few brittle fragments of bark resting on the ledge amongst a deposit of dry dust, but picture in their place a candle, a book, an ornament perhaps. The floor is also dry and powdery, and surprisingly wide: there is ample room to lie down should one choose to; room, in fact, for a bed. But these ideas of habitation are more than fantasy, and there is sense in speculating how one might indeed make this hollow a home. It is believed this cavernous chestnut was once occupied in such a way, many decades ago, by a charcoal maker of Montseny's mountain forest. Little is recorded of his tenancy in the tree but its substantial girth is rumoured to have accommodated a table and chairs during its occupation. Castanyer d'en Cuc, as the chestnut is known (castanyer being Catalan for 'chestnut tree'), is a fine example of an ageing sweet chestnut and the largest of its kind in Catalonia. It stands rather majestically on the edge of a mountain pass – a foothill of the Montseny Massif situated within miles of wild broadleaf forest.

Castanyer d'en Cuc

The massif forms part of Montseny Natural Park in the *comarca*, or county, of Vallès Oriental, known for its dramatic landscape and as a biodiverse melting pot of the Mediterranean and Central European ecoregions.

Reaching Castanyer d'en Cuc on foot is something of a test for the quad muscles. It is by no means inaccessible, as much of the journey follows a well-established track, but the hour-and-a-half hike is entirely uphill; you need to *really* want to see this tree before embarking on the pilgrimage. Fortunately I very much did, as the chestnut is one of my reasons for visiting Catalonia, and the heady walk up through the southern massif is in every way worth the exertion. For one thing the setting is stunning; views all around are of deep gullied valleys, clustered rocky peaks and a hazy Mediterranean sky that tints each steep on the horizon a different shade of blue. The forests clinging tightly to this sea swell of a landscape are a deep green produced by the dark leaves of holm oak. As the evergreen is the dominant species in the area Montseny's panoramas remain much the same through summer and winter; the latter only really diversifying with the arrival of snow. Temperatures certainly drop below freezing here during winter, confusing the senses somewhat given the arid appearance of the region's plant life. In fact, had I not assured myself of the chestnut's location prior to coming, my sprits may well have been dampened at the beginning of the walk. At the outset the resident flora is so very Mediterranean; all Spanish broom and *Euphorbia*, with rock-rose and wild lavender creeping out from under loose stone. But it is the elevation further up that produces the deciduous trees, where the soil is rich and cool and, occasionally, damp underfoot. Here chestnuts are accompanied by hazel and sessile oak, and around their roots sprout hellebores and scrambling, rampant bramble. Dirt becomes mud, in layman's terms, and the higher the path takes you the less arid the surroundings become. Being January, *Helleborus foetidus* (stinking hellebore) is one of the few wild flowers on show, its acid green petals flopping like silk over the splayed fingers of its palmate leaves. The many other flowers that will burst into life later in spring are identifiable now only by their tightly rosetted clumps. They huddle low to the ground, waiting patiently for the arrival of warmer weather, dormant as the butterflies whose larvae hunker motionless within them. Among those that are identifiable – as in this stage they are so similar to one another – I recognize hawkbit, sea holly and oxeye daisy, reminding me to spend more time getting to grips with the foliar intricacies of herbaceous wild flowers.

Even if the plants and views along Montseny's south-western range were somewhat less engrossing in character, one cannot meet Castanyer d'en Cuc itself without a sense of utter wonderment. The tree does what old chestnuts

do best; it curves and gnarls and balloons out at the base, its
senior bark fluted with deeply cut rivets. Chestnuts in effect
live two lives; similar to common ivy they begin in one form
and end in another. Just as ivy abandons its maple-like,
climbing leaf for the rounder, heart-shaped outline of
its mature shrub form, so do the linear trunks of young

chestnuts give way to twisted, spiralling bark. Though this effect is an endearing quality of noble chestnuts, Castanyer d'en Cuc itself is actually quite grotesque, not least for its footing, which appears as though melting into the earth – a globular, fissured mass reminiscent of slow-moving lava. Its shape pours over two levels of ground, the surrounding soil presumably having shifted gradually downslope during the tree's many years on the hillside. There is no trunk to speak of – that single distinguishing feature by which we classify trees as trees – instead, wandering out from the peculiar central bulk, is a collection of aimless limbs that thrust into the air like tentacles belonging to some urchin of the seabed. They vary in breadth and have a weightless quality, as if lacking any particular purposeful direction. Circling the tree's immense base (listed in some accounts as 40ft/12m in radius) I count 13 of these main stems, most of which are attached, like a coppice stool, to the lower section of the trunk. Some of the branches are comparatively young and display the thin sheen of new bark. Two stout segments remain of the original trunk; having died off at the top both now stick out as cylindrical chimneys. And at the very bottom, protruding from under the folds of molten bark, brand new shoots appear, half an inch (1cm) thick, carrying the chestnut's signature fat, red buds that are every bit as healthy and exuberant as a nursery-grown sapling. These many components together – old, ageing and new – form an organism emblematic of struggle and survival: it is a tree wholly un-treelike and devoid of conformity, a character sculpted by the demands of its challenging environment. Fire, most likely caused by lightning strike, has altered its torso dramatically, cutting the plant down and hollowing out its centre. Drought, in turn, is most likely responsible for the waning limbs, especially given the tree's exposed positioning at the mercy of furious wind and a blazing sun. The chestnut may even have been pollarded at one time too, or cut from at intervals by local foresters. But this is speculation; who, at a glance, can really account for the many adversities faced by a specimen of this maturity and isolated existence? We merely see the contortions that have resulted from their infliction.

As I stand admiring the chestnut, I find myself drawn to that incongruous church window at the centre of its trunk. I'm reminded of so many trees with similar cavities; the oak in North London's Waterlow Park, for example, in which children play hide and seek, or the enormous cedars I once met on an estate in Devon whose holes had been filled up with tar. Ageing trees are so often treated in this way amid fears for their lasting stability; packing the exposed centre (usually with concrete) is a measure intended to keep them upright and alive. Though this is relatively harmless to the tree itself it is generally a cosmetic measure; with all ventricle tissue located on their outer layers trees can live on quite happily lacking any substance to their core. There is no better testament to this than Chêne Chapelle in France, an oak over 800 years old that has withstood similar excavation by lightning. The immense tree stands in a small farming village outside the northern city of Rouen, and has become a destination for religious pilgrimage ever since a chapel was erected in the entrance of its hollow. The lightning strike took place sometime in the 17th century and, taken as a sign of divine intervention, prompted a local abbot to construct within it a shrine to the Virgin Mary. Today, Catholic mass is still conducted twice a year from within the cavity, which is fitted with a paved brick floor, wood-panelled interior and an ornate altar against its back wall. There is even a second chamber higher up known as the hermit's room, accessed via a staircase that spirals around the outside of the resilient trunk. Parts of the exterior are clad in wooden shingles (in lieu of interior concrete), to cover the natural crevices and holes that have broken through over the years, a kind of curious domesticity added to an already adulterated organism.

But give me Castanyer d'en Cuc over Chêne Chapelle any day. If I were a hermit I'd far rather take the chestnut's empty, earthy shell and the quiet of its hillside forest, even if it meant paying the penance of that long uphill slog. Give me a few leaky holes in the trunk if they let in the winter sky, and leave the walls unclad if they present a nook for the odd nesting bird. This is a chapel as nature intended; a tree far more fitting for a hermit, religious or otherwise.

Pannage in the New Forest

D riving through Surrey I pass the sign for Willey Mill and realize I am on a familiar road. I ought to have noticed sooner but my internal compass has been subdued by the dawn start and an unhealthy reliance on satnavs. I came this way to Hampshire at the beginning of last year, back in March with the roadside trees still bare of leaf and white with the first blossom of spring. Spring, in fact, was what I had come to see, or 'find', as had been the case in this instance. It was the poet Edward Thomas's road, leading from Farnham to Alton, past little villages and along the winding River Wey. Sure enough, a succession of road signs promptly produce 'Bentley', 'Froyle' and 'Holybourne': all points at which I'd stopped along my journey tracing Thomas's footsteps for an article marking a century since his death. In 1913 Edward Thomas had set out from South London to seek the signs of spring in the western counties of England. He mapped the season's arrival in their fields and woodlands, in village churchyards and streams; all senses tuned for the sight of opening wild flowers and the songs of nesting birds. His observations were recorded and later published under one of the loveliest titles in the nature writing anthology, *In Pursuit of Spring*, a book written at the outbreak of a war that would rob Britain of a generation of poets and inspire the poems with which they were later immortalized. While Wilfred Owen lamented trench warfare and Rupert Brooke confronted death with stoic, resigned verse, Thomas's poetry continued to elevate the natural world, recalling the nuances of his cherished home country hills and hedgerows. Four years before his tragic death in France during the First World War, Thomas's book commission had had him nervously excited, anxious to set forth from London and reach 'the nightingale's song, the apple blossom, the perfume of sunny earth.' Shafts of sunlight would flood through the windows of inner city rooms, falsely heralding the arrival of spring, and memories were triggered of a familiar and beloved countryside that lay just beyond, beckoning him to begin his journey. 'I knew how the first blackbird was whistling in the broad oak,' Thomas wrote at the outset of his pilgrimage, 'and, further away – some very far away – many thrushes were singing in the chill, under the pale light fitly reflected by the faces of earliest primroses.'[1]

My pursuit of Edward Thomas, along the A31 from Farnham, had been an immeasurably happy one. Captivated as he had been by the unveiling season, I found exhilaration in seeking out celandines on the hillsides of Hampshire and the minute petals of moschatel, spread low through its damp woodlands. With *In Pursuit of Spring* as my guidebook I listened for the birds Thomas had heard and stalked the plants he had recorded – each a footprint in the mud marking his course from village to village. These encounters were made all the more enthralling by the knowledge that they had been witnessed too by him, a hundred years ago. Returning to Thomas's path this morning I feel again that excitement of escaping the city. London certainly isn't short of green spaces but a claustrophobia can nonetheless build, fuelled by a need to 'get out' and to rebalance the senses with unobstructed vegetation. I'll soon be turning from Thomas's road and instead continuing further south, headed in the direction of one of England's most treasured and protected rural landscapes. My getaway today is to the New Forest, a natural refuge popular among bolting Londoners. In further contrast to my trip at the beginning of the year, on this occasion I am in pursuit of autumn, seeking in the medieval woods of southern England a particular occurrence indicative of the season.

Reaching Winchester I join the M3 and continue on towards the south-west. The New Forest occupies the lower reaches of Hampshire, situated more or less in the centre of the mainland coastline, facing the Isle of Wight. As a National Park it covers over 200 sq. miles (500sq. km), incorporating numerous towns and small villages, the 'capital', as it were, being Lyndhurst. This seemed the obvious place to begin my autumnal assignment. It is a smallish town, yet a busy one: it acts as the administrative hub of the forest and therefore hosts a crossing point for the significant roads serving the region. A little way before Lyndhurst however, at the end of the M27 near Cadnam, it is already evident that I have entered the New Forest. The roadside trees that had been steadily gaining mass now rise high above the motorway, obscuring from vision any landscape beyond. But the true spectacle of the area becomes obvious as I change to smaller A-roads leading off towards Lyndhurst. My route takes me through wide open heathland and beneath

tracts of arching beech and oak trees; traffic all but peters out
and lengthier parcels of ornamental woodland come into view.
It's now late October and the road verges are already flanked
with fallen leaves; their colours every imaginable variant in
the spectrum of brown. The roads are bright and wonderfully
empty, lit by an intense morning sunshine broken upon
horizontal tree boughs.

The word 'forest' has not always implied an area made up of trees. It is a term that has shifted over centuries from one understanding to another. The modern application would rarely conjure in the imagination anything other than a grouping of trees, however for much of the Middle Ages the term was associated more directly with something similar to 'pasture'; a landmass on which animals were grazed and sustained, comprising numerous contrasting features. The topography of a forest could therefore range anywhere from open heathland to dense woodland, and livestock were as synonymous with the word 'forest' as trees. As such, wooded areas were once measured according to the quantity of animals they could support, rather than by the types of trees therein.

This association of forest with animals changed little following the Norman Invasion of 1066; however, what *was* altered under William the Conqueror was the accessibility to certain forests for the peasantry already resident there. While on the throne the new king siphoned off rural regions as 'royal hunting grounds': areas of forest now belonging to the Crown in which nobility had exclusive right of use. Of these designated royal forests, the New Forest is perhaps one of the better known. Its name in fact derives from the Latin, *nova foresta*, meaning 'new hunting ground'. Royal forests were governed under strict forest law, designated as areas in which the King's animals (notably deer, boar and hare) were managed and protected. The appointment of these numerous royal grounds had significant implications for those whose way of life had until then greatly depended on them. Where once peasantry had been free to run livestock, collect timber for fuel and hunt game to feed their families, now such acts were prohibited. Anyone caught either hunting the king's deer or disturbing the understorey (which could include anything from removing fallen wood to fencing-off crops) were subject to heavy penalties. Typically, punishment for these offences, particularly those instigated under William's successor, William Rufus, were severe; punishable in many cases by death.

It is no surprise that such stringencies enforced upon the peasants' use of land were deeply unpopular. As royal forests steadily increased in size, so did it become progressively challenging for residents to work their plots effectively. In 1217, following 150 years of strict royal governing, forest law was at last revised, and two years after the Magna Carta – and considered by many as an adjoining document to it – the Charter of the Forest was drawn up, returning to the commoner certain rights and privileges that had been prohibited under Norman aristocracy. Those reinstated included the foraging of firewood, the grazing of ponies and the custom of pannage – allowing pigs out to graze on fallen nuts during the autumn season. My visit to the New Forest today happens to fall almost 800 years to the day since the charter was instigated. Issued early in November, the forest of the 13th century will have in many areas looked similar to the way it does now; deciduous leaves falling with splendid colour and the fruit of oaks, beeches and chestnuts strewn among their litter. Whether or not commoners in the New Forest of November 1217 took upon their reappointed rights immediately and turned their swine loose under the trees to fatten on acorns, nuts and beech mast, is unclear; however, the privilege of pannage has continued ever since, and it remains customary that one may, while wandering through the New Forest at this time of year, encounter free-roaming pigs rootling through the undergrowth.

This novel activity forms the object of my autumnal pursuit today, as do the chestnuts with which the season, and pannage, is closely associated. As noted earlier, autumn reveals in the woods of the New Forest as a multitude of attractive browns. Among them, however, chestnut is perhaps the most captivating of all; a colour denoted singularly by the nut, and iconic, therefore, of the season. Edward Thomas's mastery of observation might surpass others in characterising spring, but for evocations of autumn one might turn instead to Thomas Hardy. Passages in *The Woodlanders*, for example, describe the season falling upon Hardy's fictional Wessex with poignant accuracy; he writes of orchards 'lustrous with the reds of apple-crops, berries, and foliage,' and of these being, 'intensified by the gilding of the declining sun.' In his conjuring of the season Hardy injects a physicality, recounting that hedges 'bowed with haws and blackberries', acorns 'cracked' under foot, and chestnut husks burst, 'exposing their auburn contents ....'[2] As in Hardy's Wessex, sweet chestnuts crop heavily in the New Forest. However, by no means do the trees themselves proliferate; they tend to grow sporadically and only in certain locations within the forest perimeter. As a whole, the New Forest is an area of great diversity, boasting a wide range of differing habitats and an impressive variety of trees. There are stretches of wood, for example, championing enormous Douglas-firs, and in contrast swathes of downy birch can be found on boggier ground in the forest. Around each corner there are stocky stands of oak and beech yet in many places these intermix with evergreens like yew and holly. Therefore locating the New Forest's chestnuts isn't particularly straightforward. But what better excuse to escape the city, I think to myself as I arrive in Lyndhurst, than to go looking for chestnut trees and pigs in the footprint of a medieval forest.

I will happily admit that scouting for pigs in woodland is not within my field of expertise; scouting for chestnuts perhaps a little more so. In locating the latter, therefore, I am leaning upon my own arboreal decipherment, whereas for the pigs I have come prepared. Help was sought by way of the Verderers of the Forest; a council under whose jurisdiction the free-ranging pigs lie. The verderers are an appointed court within the New Forest made up of representatives from a range of governing bodies, such as the Department for Environment, Food and Rural Affairs (Defra), the Forestry Commission and Natural England. In addition, their administration includes elected members local to the New Forest, and a representative appointed by the monarchy known as the Official Verderer. The verderers are charged with protecting the landscape of the New Forest and overseeing the rights of commoners within it; they ensure that these are being carried out appropriately and in accordance with the laws of the forest.

The presence of such a council in the New Forest likely dates back to the 13th century, to sometime around the arrival of the Forest Charter. However, their official role in overseeing commoners' rights came into effect following an act of parliament in 1877. Rights of Common are attached to land situated within the forest; so it is to the verderers that one turns when purchasing a property inclusive of one or many such rights, in order that these may be lawfully understood and exercised. For example, a New Forest property may come with estovers rights – an allowance of wood fuel in the form of firewood. This is collected from the understorey of the wood or pre-cut and distributed (in modern times) by the Forestry Commission. As for the right of autumn pannage, the verderers, with the assistance of the Forestry Commission, decide upon the beginning and end dates each year, electing the period during which commoners' pigs may be let loose into the forest.

Having served as clerk to the Verderers Court for over 20 years, Sue Westwood has a great deal of experience with the ins and outs of pannage. I contacted Sue ahead of my visit to enquire after the New Forest's pigs and where I'd most likely find them. She kindly responded with a list of suitable locations and areas which might prove useful. 'Chances are you'll find them,' she had said on the phone, '... unless you're actively looking for them.' Sat in the car at Lyndhurst I scan through Sue's list on which I've circled the names, Burley, Bolderwood, Bramshaw and Anderwood. From Lyndhurst, Anderwood is the nearest, and a map reveals its location as halfway along the road to Burley. I should say at this point that although my choices were selected purely on the basis of nearest first, both Anderwood and Burley proved to be very fortunate locations. Had I not explored Anderwood I might have spent the rest of the day searching in vain for chestnut trees, and if not Burley then the same for pigs.

Coming off Lyndhurst Road and up through tall conifers into a small car park clearing at Anderwood, I find myself scanning the vegetation for any sign of movement. I have pigs on the brain and am determined to spot them, however elusive they may be. The Forestry Commission, who maintain all aspects of forestry in the New Forest, have encircled the parking area with a low fence; walkers are advised that they may picnic in this spot before entering the wood through a designated wooden gate. I park up, pass the gate and head eagerly into the deciduous trees ahead, eyes still fixed upon the lower storey. Distracted by my pursuit I am oblivious to having stumbled straight in among a stand of fruiting chestnuts. In surprise I look up to see the elongated, finely toothed leaves indicative of *Castanea sativa*: what little remain on the branches are a contrasting mix of deep green and light brown, curving gently downwards at the edges.

The trees, growing densely here, become perceivably unique in character; they are so different as a group to beech or oak, and I check myself for not having noticed straight away. Their forms seem unruly and distinctly uneven, with limbs jutting in atypical form from wayward, leaning trunks. I'm familiar with the coppiced chestnut woods of East and West Sussex, where management has effected much straighter growth habits in the multi-trunked trees. However, these single-stemmed chestnuts are quite the opposite: far less uniform and quite wild. I check the ground around my feet, pushing a mat of leaves to one side, and find an abundance of brilliant green seedcases. Their spiked exteriors match, in colour and form, the great clumps of green star moss that also inhabit the base of the stand. Beside the seedcases lie scattered chestnuts, reflecting sunlight with a sheen of woodland floor damp. Exercising my basic peasant's right of forage (though this is not strictly a commoner's privilege) I collect up a handful of nuts, and place them in my jacket pocket. Through the fabric I can feel their permeating cold, the cold of the forest sleeping dormant in its seeds. The profusion of chestnuts lying beneath Anderwood's trees form quite an extraordinary bounty, a concentrated number greater than any I've really seen before. I wonder though if this might suggest a certain absence of pigs, and that the secret is not yet out among the forest's roaming swine.

With Anderwood explored and its chestnuts found, I return to the car and carry on to Burley. Burley, I later discover, is a sweet little village; in every way fitting with the image of rural England. However, I'm pulled up short before reaching the village outskirts on account of a large sow and accompanying piglet emerging suddenly from the roadside scrub. I park and jog back to the spot, listening for a rustle to give away their position. A short distance from the road the bulky pink sow reveals herself again under a stand of young oaks, and on hearing my approach she comes waddling over, with the piglet trotting behind. Evident from their confident approach these pigs are clearly used to people. Nevertheless they are a peculiar sight as they wander freely by the roadside, as if broken loose from a pen. Failing to produce anything of edible appeal, the pair quickly lose interest in me and go on nosing the ground in search of fallen food. I watch as they jolly, heads to the ground, focussed on the task at hand.

But there is more to this novel sight than the execution of an ancient autumnal tradition. The commoners' right of pannage places an animal of the farmyard back in its rightful domain, into the wooded context of its ancestral heritage. The wild boar – the forbearer of all domestic pigs – is every bit a creature of the woods, though in Britain this is a fact often forgotten. Our most frequent encounters are with pink-skinned 'porkies' lying in the mud of farmyard fields. To look at a European wild boar however, there's no mistaking how suited a creature it is for life in a woodland terrain. There's the enormous snout for rooting – and sniffing – through leaf litter; the thick set shoulders for shifting heavy logs; hair for winter warmth (shortened or absent in domestic pigs); hooves for balance on uneven ground; and those sharp tusks and teeth indicative of boar, capable of unearthing topsoil and biting clear through the roots of trees. The boar's nocturnal habits renders eyesight of little value, which, much like a badger, accounts for its little eyes and poor vision. Hearing, therefore, takes up the sensory slack, resulting in hefty, erect ears. And if all this wasn't evidence enough, wild boar piglets ('boarlets') are born with a camouflaged coat streaked in light browns, matching the woodland floor with surprising accuracy. Sadly though, unlike most of Europe, Britain's wild boar population was pushed to extinction by the 17th century through excessive hunting and diminished habitat. In recent years however, a small number have returned, particularly in the south-east of England. They are assumed to be escapees from farms and wildlife parks, and there is evidence that their number is increasing.

In 1998, Defra launched an official investigation to ascertain the long-term effects of this reintroduction to the UK. The resulting report confessed a lack of pre-existing Anglo-hog literature with which to inform its conclusions, and therefore takes the shape of an in-depth risk assessment, seeming to pose more questions than it answers. Twenty years on and little has been decided as to the future of wild boars in Britain. Meanwhile, beavers have returned and are on the up in Devon and Scotland, and the lynx is forever in discussion. Perhaps though, as an animal still prevalent in mainland Europe, the wild boar is a good example of the British Isles being simply too small, or at the very least, too deforested, to accommodate the beasts it once did. Time will certainly tell.

The day after my escape to the New Forest, back in London, I remember my pocketful of chestnuts and fetch them from my jacket. I roast a few in the oven, wrap them in a paper bag, and take it with me into town. On the train ride in I put my face into the bag, like a greedy pig, and breathe in the warm, autumn woodland. The smell is always better than the taste; sweeter and richer, redolent of earthy forest. But while the chestnuts are still hot and easy to peel I devour a handful. They are a little dose of countryside; a tonic taken neat, a quick fix of the forest for another day in the city.

Poplar

Poplars belong to the willow family. In mainland Europe it is the fastigiate Lombardy poplar (*Populus nigra* 'Italica'), those tightly-branched, columnar trees lining the hills and farmland boundaries of Italy and France, which we associate most readily with the name poplar. The Lombardy poplar is a cultivar, one of the many varieties of Eurasia's black poplar, *P. nigra*, widely cultivated from cuttings. Across the Atlantic, however, the genus *Populus* also includes the cottonwoods – the name given to poplar species native to North America, including *P. deltoides*, *P. fremontii* and *P. trichocarpa*. Cottonwoods cling closely to the sides of rivers and creeks where, during spring and early summer, their seed is dispersed by the wind in clouds of soft, white fluff; a feature that has led to their elucidatory common name.

Also included in *Populus* is the aspen, the American species (*P. tremuloides*) and the European (*P. tremula*). If the cottonwood has the fluffy seeds and the Lombardy has the figure, then aspens are characterized by an attractive and compelling foliage. Many deciduous trees are exalted for their stunning autumn leaves but those of aspen are both colourful (yellow-orange-gold) and audible. Their propensity to flap and twist on their stalks is such that only the slightest of winds, passing through the canopy, will raise a remarkable clamour. On account of this, the species bear the respective common names of trembling and quaking aspen.

Aspen (*Populus tremula*)

Autumn leaves cascade onto the motorway towards Oxfordshire, whirling in the crosswinds between streaming cars and the blusters of Hurricane Ophelia. On the tarmac they flop and flounder like shoals of stranded fish, thrown into bundles held down by the driving gale. Ophelia is making headlines today for whipping up Saharan sand, carrying it over the continent and staining the sun red. Back in London an apocalyptic, carmine sphere glows on the skyline, its light refracted by the high-sailing particles swept in from North Africa. Stranger still, today is 30 years – to the day – since the Great Storm of 1987 tore through Europe, devastating southern Britain and knocking down an astonishing 15 million trees. Losses in the landscape on that scale had not occurred for centuries; a great flattening of woods right across the south-east from the Chilterns to Kent, where in places up to 97 per cent of the canopy was razed to the ground. Beech, with their hefty limbs and shallow roots, were hit the hardest, particularly those sat lightly on England's downland chalk. But among the blanket of wreckage were trees of individual note too: many of London's landmark planes and, famously, six of the iconic oaks in the Kent town of Sevenoaks. The sense of mourning in the storm's wake rang in a new era of tree appreciation and preservation. As if for the first time, the trees of London's streets were seen as impermanent, ephemeral objects, landmarks that could topple, become diseased, even die.

On the occasion of the 30-year anniversary, reminders of 1987's immense storm have circulated in the media this week. The coincidental arrival of Ophelia's strong winds and dust-laden sun, however, have contributed a further, illustrative commemoration. In the spirit of arboricultural devastation and loss, it seems equally fitting that today should be the day I visit Binsey in Oxfordshire. The village is the setting of Gerard Manley Hopkins' poem, 'Binsey Poplars': a lament for a row of aspens that had once lined the nearby banks of the River Thames. Unlike the cherished trees toppled during the Great Storm however, Hopkins' poplars were brought to the ground, seemingly abruptly, by a human hand. 'My aspens dear, whose airy cages quelled,' opens the late 19th-century lament, '... All felled, felled, are all felled.' As an elegy for an artefact of the natural world, the poem's strength is its deeply personal voice and incensed tone; '*My* aspens dear ... Not spared, *not one*.' It conveys a sense of ownership suggesting that landscapes, unlike buildings, transcend proprietary rights. They are a visual property also, as are their components, possessed by those who encounter them regularly. Having studied and

Loss in the Landscape

later been employed in central Oxford, it is likely that Hopkins knew the poplars well and their disappearance from a familiar landscape was conceivably unsettling.

Whether or not the trees belonged to Hopkins however, poplars – indeed aspens – belong indisputably on the riverbank – a tree as synonymous with water as with land. Similar to willow and alder, poplars typically thrive in ground adjacent to streams, rivers and lakes; riparian terrain suited to certain and distinct plant communities. Riverine trees like these are united in their exploitation of a watery habitat, distinguishable by their methods of propagation and dispersal. Their

seeds therefore float and will germinate easily in damp soil, or, in the case of crack willow (*Salix fragilis*), branches break cleanly and can root like a cutting into mudbanks downstream. The poplars Hopkins eulogizes are likely to have been European aspen, *Populus tremula*, a species of poplar native to Britain that favours open sites and unobstructed sunlight. With Binsey situated beside wide water meadow and cattle fields, ('and wind-wandering weed-winding bank'), it is easy to imagine the poplars' removal as being a visible, if not tragic loss to the landscape.

Having arrived at Binsey I follow the village footpath to the Thames and amble south-east along the bank in the direction of Oxford. The persistent winds seem to have deterred visitors to this stretch of the water, save for a pair of stoic fishermen casting lures at hidden bream. I'm soon walking below willow and broadly trunked limes; poplars too, in pre-autumnal shades of green turning yellow. Many leaves have already unhitched and are collected on the river, swirled – as in some windswept corner – where the current turns back on itself and eddies beside the bank. Most are blotched with black leaf spot, a fungal infection commonly found on European aspen. Those remaining on the tree however – still the vast majority – flutter noisily as they spin on flat petioles, producing the 'trembling', rustling sound for which the tree is famed. To stand beneath the larger of them is to be engulfed in the uproar: a hiss like the sound of continuous waves breaking on a shoreline, or the wing-rubbing of a million crickets on a tropical evening. Only the quacking calls of nearby jackdaws are audible over the tumult in the framework of branches – the 'airy cages', described by Hopkins, that 'Quelled or quenched in leaves the leaping sun'.

A line often quoted from 'Binsey Poplars' – the conservationist's slogan, if you like – comes towards the middle of the poem:

> O if we but knew what we do
> When we delve or hew –
> Hack and rack the growing green!

For Hopkins it is the irrevocable nature of the violent act upon the poplars that becomes the object of his rage and the wider message of his poem. It is the interference with Nature; damage to a 'Sweet especial rural scene' that, once gone, cannot be replaced. 'After-comers cannot guess the beauty been', he solemnly declares; the splendour will be forgotten. We're given little indication as to why the aspens were felled, though Hopkins skips past the ambiguity by observing yet another theme of conservational concern. Shifting in voice so as not to exonerate himself from humanity's habits of destruction, he adds, '... even where we mean to mend her we end her,' implying that no good can come of any intrusion upon – or dominance over – what Nature has created, even with the best of intentions.

In the aftermath of the Great Storm of 1987, frenzied clearing and replanting took place at sites floored by the battering winds. Partly due to the sheer scale of the task at hand however, many areas were designated as 'exclusion zones'; places where trees were left uprooted and horizontal, lying where they fell. Some years later, encouraging reports began surfacing regarding natural regeneration in these untidied areas; examples such as at Scords Wood and Knole Park in Kent, where woodland appeared to be restoring itself, springing new life – and new trees – from the disturbed and exposed ground. RHS Wisley lost many of its significant trees, yet the reopened canopy encouraged diverse wild flowers and fungi to bloom, previously restrained by a lack of light. It seemed that efforts to manually revitalize the landscape – through clearance of dead wood and the establishment of new saplings – were being far exceeded by Nature's own coping mechanisms. Furthermore, witnessing these developments led people to conceive that natural catastrophes could in fact impact positively upon wooded environments. As the British nature writer Richard Mabey once reflected, the storm, like so many before it, demonstrated 'that natural disturbances were an entirely normal and well-tolerated part of a woodland's experience.'[1]

From further down the river, looking back at the rustling poplar, I'm reminded of my own experiences with losing trees. I recall the period when a dozen false acacias were due to be cut from a garden which, for many years, I had looked after as head gardener. Under new ownership the garden was being radically redeveloped and given a more formal, simplified design. In the days before my departure a team of landscape contractors set about dismantling what had been a wonderfully wild oasis; shrubs were pulled up and borders emptied, even the meadows were reseeded with turf grass. Seeing the planting unearthed like this was hard to stomach, as was the filling in of the little wildlife pond beside my old shed, but hardest of all was the arrival of yellow 'X's spray-painted on the bark of trees I'd grown so fond of. The acacias, some of which were 80-foot monsters, bore commanding boughs with glorious, deep fissures, frequented daily by nuthatches and woodpeckers. In spring their leguminous flowers lit up the garden like no other tree, dangling hundreds of creamy white chandeliers over the woodland glade. To think of them felled brought a sense of sincere foreboding: herbaceous perennials are temporary, a pond can be replaced, but trees are incalculably unique. I wonder now if I'd have felt differently were the trees instead brought down by a surging storm. Would their loss have been easier to accept if

inflicted by natural causes? Probably yes, though I'd still have lamented their sudden end. Our affections for trees as individuals go beyond ecological reasoning, which, ultimately, is what makes us human. The more we understand of the natural world the better we can preserve it, but there will always be room for a little sentimentality.

The dining room of Boardman's River Lodge and Grill offers, as the name might suggest, an appealing view across Oregon's Columbia River. Rising in the Rocky Mountains of eastern British Columbia, the enormous body of water passes the town of Boardman on its journey to the Pacific Ocean, fed, at intervals, by numerous tributaries as it pushes out west. The window by my sunlit breakfast table frames a vista of expansive flat water, edged at the far end by a subtle rise of sandy coloured hills. Like so many of the great North American rivers, the Columbia denotes a state border, and here divides northern Oregon from its neighbouring Washington, to which those hills belong. On the near bank a handful of plovers scuttle across collected river stone and a few gulls climb gently onto the breeze above. Whether by dust or industrial impurity, the morning sun is defused into a washed-out haze and distributes its light evenly across the view. Having arrived at the hotel late last night at the darkened end of a 450-mile drive, the placid scene is wonderfully restorative; I sit happily bewildered by its scale and simplicity, still floating somewhat from the journey. Looking out at the river one might be fooled into thinking it a wide lake; the sheer volume and apparent stillness play easily with the mind. A fish rises some way out from the bank and the responding ripples disseminate unhurried. Interstate 84, a highway running up to the Columbia from Utah, joins the river here at Boardman, and ships its traffic close to the water 160 miles on to Portland at the cool, misty end of the country. But the forests and waterfalls of that region seem a very long way away, given their stark contrast with the landscape and climate of Boardman's surrounding prairie.

Oregon Tree Farm

Typical of the towns found dotted along the interstate this side of the Cascade Mountains, and indeed those of the prairies further south-east, the town of Boardman has evolved in line with large-scale industry. Driving the I-84 across Oregon from Idaho (as I did a few years ago) one encounters some of the vast infrastructure of the state's key exports. Generally speaking these are a mix of agriculture and construction, the former being particularly prevalent now in the grasslands of the Columbia Basin. As a member of the River Lodge's staff tells me later that morning, 'In Boardman we got potatoes, cows and coal ... and not a lot else.' Although to my delight and relief he hastens to add, 'Oh, and the tree farm of course.' The tree farm is what has drawn me to this northern stretch of Oregon, a man-made 'forest' of gargantuan scale, dedicated to the cropping of distinct hybrid poplar trees. As with the majority of the industry set up in this area, the farm is inextricably bound with the Columbia River, dependent upon it for both irrigation and electrical power. In fact, the Columbia is the lifeblood of this terrain and that much further beyond; a natural resource so well tapped by colossal dams that, according to a recent article in *The New York Times*, along with its tributaries the river supplies half the nation's hydropower electricity.[2] This excessive damming is the reason why the water moves so slowly past Boardman, and its sedentary fish, hindered on their migrational path upstream, are one of the controversies of the river's drastic alteration.

Siting over 24,000 acres (9,712ha) of forestry in arid Oregon prairie would seem inconceivable were it not for Columbia's great dams and their notable history. The area around Boardman averages around 8in (20cm) of rainfall a year; often less than an inch during the summer months. Therefore, the millions of poplar trees grown at Boardman Tree Farm rely entirely on water pumped from the river. It is, in fact, one of the biggest drip irrigation systems in America, fed via a network of massive underground pipes. As water is directed through progressively smaller irrigation lines, GreenWood Resources Inc., who own the farm, balance watering volume delicately across their grid-system fields, making sure that each tree receives an exact amount for optimum growth. Talking with Austin Himes, Harvest Manager at GreenWood Resources Inc., it's clear that a great deal of monitoring is required. Close attention is paid to local weather station forecasts and moisture levels in the soil. Austin explains that much of this demand is down to the trees themselves, hybridized poplars selected for geographical suitability, lumber quality and growth rate. 'GreenWood Resources has one of the world's leading hybrid poplar breeding programs,' he tells me. 'We hybridize species of *Populus* from the 'cottonwood' section (*P. deltoides*, *P. fremontii* and *P. nigra*), conducting extensive field trials to achieve specific attributes.' Speedy growth is certainly a cottonwood attribute. British tree expert and author, Hugh Johnson, noted that a cottonwood tree planted on Mississippi bottomland was recorded as

having grown to a towering 98ft (30m) in 11 years.[3] 'Hybrid poplar is the fastest growing tree species in the temperate zone,' Austin affirms; an attractive quality for a tree intended for regular, large-scale cropping.

As a farmed tree, hybrid poplars have a great deal to offer by way of end product. Almost all aspects of Boardman's trees are processed, mostly onsite, into various profitable commodities. Their lumber is used in the making of furniture, winter sports equipment and, on a much smaller scale, pencils. Core lumber is processed for hardwood veneer and resulting wood chip is compressed for biomass systems (boiler fuel, for example). Leftover bark and sawdust is sent to farms as bedding and agrarian top-mulch. Pulp for paper, says Austin, is another key use for the wood. 'The fibre from hybrid poplar is short, white and bright, which are desirable characteristics for some types of paper production. The wood takes stain easily, it is very light weight and has reasonable strength properties.' Furthermore, it is again the loose ends – the excess lumber and scraps of wood chip – that are converted to paper pulp, extending the versatility of the poplar's economic value.

The Boardman area was originally chosen because of its proximity to the Columbia River, I-84, and paper manufacturing facilities. This existing infrastructure insured economical transportation of goods with reasonable access to international ports and inexpensive electricity. A drive to 'settle' the area in the early 1900s led to the erection of gigantic concrete structures on the Columbia, the Grand Coulee and Bonneville dams being of notable prominence. Beginning construction in 1933, the Grand Coulee remains today one of the largest man-made structures ever built. Its immense girth comprises close to 12 million yards (11 million metres) of concrete, trailing a reservoir 150 miles behind it. The original purpose of the damming, led by the Government's Bureau of Reclamation, had its roots in social development. It was intended that over 1,500sq. miles (4,000 sq. km) of profitable independent farms and homesteaders in the Columbia Basin would be the beneficiaries of the enormous project, funding their newly irrigated fields through affordable government repayment schemes. It was one of the heralded solutions to the Great Depression, an antidote to the growing dust bowl at the centre of the country. However, by the time the Grand Coulee was finally completed, hydropower took precedence over irrigation, a priority enforced by the demands of an escalating Second World War. It wasn't until the 1950s that water began to flow in the direction of irrigation, and sadly by this point the socialist ideal had shifted, through blurred lines and escalating costs, so that larger corporate companies had become its chief recipients. Furthermore, the substantial damming along the Columbia radically impacted upon its resident and transient wildlife (such as migrating salmon) and the Native Americans whose way of life once intertwined with them. I reference this short history firstly to expand upon the reasons for locating a thirsty commercial tree farm in hot, prairie Oregon. But secondly to illustrate something of the context that travelled with me into the man-made woods at Boardman. The river dams introduce America's relationship with its immeasurably challenging, virgin terrain, and their story will hopefully embellish any description I might attempt of a 'gigantic tree farm in the desert' with a backdrop of equally artificial yet bold constructions erected in the wide and 'wild' West. As with much of America's landscape, the Columbia Basin presents what the author Richard Mabey called 'a scrap of real wilderness', a landscape retaining its rugged, untamed natural heritage. As he later concludes; 'America has never entirely swallowed the British belief that every last inch of the land must be managed. Nature is taken *seriously* there.'[4] To this end, the structures of industrial enterprise sit isolated in the prairies of Oregon and Washington, 'subverted', you could say, by the wild country that surrounds them. Just as the concrete of Grand Coulee jars with the Columbia's rushing water, Boardman's regimented trees appear peculiar in the desert dust.

It is late afternoon when I make my way to the tree farm for the first time. Thinking it preferable to avoid the midday heat, I set out from the hotel once the sun is past its peak. Locating the trees is pretty straightforward: the front line of their many planted rows stretches up close to the interstate, forming a wall of living pillars. What can be seen from the road, however, is just a fraction of the farm's full footprint: a single plot on a giant grid; one square of the vast mosaic. Pulling off the interstate and driving with my back to it along the line of the trees, similar sized tracts of forest appear on the horizon, uniformly angular, containing specimens of uniform height. The patches of green add a missing depth to the landscape, showing its scale through distant features. As my visit to the farm has unavoidably landed on a weekend, Austin kindly provided a map of the plots, granting me permission to roam alone through the plantation. Across the map is scrawled handwritten notation, indicating locations of particularly mature stock for me to visit. It takes a moment to figure the layout, cross-referencing plot numbers with the tags that I find at the corners of each planted section. I liken the map to the grid-system of American towns: inhabited blocks derived from a simple, sensical geometry. Access roads divide the 290-tree-per-acre blocks; dirt tracks forming impressive and attractive avenues. A wide road runs through the middle of the farm, at the centre of which sits the resident sawmill charged with processing Boardman's mass of timber. I wonder if the road ought to be called 'main street', to observe its orderly function. The whole enterprise is so different from my experience of forestry in the UK, more regimented even than the coniferous copses of Scotland and Wales and dwarfing them in scale. After a ten-minute drive I reach what I hope is the indicated plot, pulling up by a procession of thickly girthed poplars. Stepping out of the car I am met with silence. There is a distinct and familiar smell of farm; a sweet manure hanging in the air. The trees seem welcoming at first – the bare trunks allowing light to filter evenly through the initial columned rows. The canopy is an array of greys and greens, its colours layered like sandstone, one on top of the other. What seems strange is the ease with which the organized wood may be compared as much with an enormous, barren warehouse as with a sentient, living forest.

Its unbroken regimentation is at once sterile and alive. Oddly, I feel hesitant to enter. Despite the wood's unnaturally organized composition, the brain cannot help but register some primal danger, interpreting the collection of trees as an uncertain entity. I recall the same feeling as a child, stood at the perimeter of a dark fir or spruce plantation, deterred by a sense of unease. Something in our ancestry clearly associates forest with danger, and I don't believe it would be difficult to hypothesize as to what. It is well recounted that the vast majority of Britain's landmass was once densely forested. However, according to the Woodland Trust, it took only until the Middle Ages for woodland cover to be reduced down to an alarming 15 per cent. The 'wildwood', as ancient forest has been dubbed, was a home to carnivorous mammals, a harbour for vagabonds and a hindrance to agricultural development. 'We have a strong affinity with this wildwood,' writes forestry expert Chris Starr, 'and it is reflected in our fables, our literature and in our deepest desires and fears.'[5] The fear, it would seem, remains in connection with trees, whether amassed as a natural forest or one fabricated by man.

With a little excitement I walk in through the first few rows. The ground is dry and rustles as I move, covered for the most part with low, seeding grass. At the base of the trees run leaking black cables, dripping quietly onto designated patches of dark, damp earth. It is still some weeks before the summer will be over and what few leaves have fallen are far from turned. I pick one up, surprised by the size of its familiar shape. The leaf's wide triangle fills the palm of my hand. It is an attractive form, domed like a broad trowel and flat along the bottom. The margin is serrated like the edge of a lime leaf, although less jagged and severe, smoothed into bumps. The lineage of these hybridized trees shows up in their veins: bright lines cutting through the dark green leaves recall those of black poplar (*P. nigra*), one arm of their intergeneric breeding. As I move further through the rows the sun lowers and begins to project an eerie orange through the wood. The trunks are marked by its Martian colour, like bottom-lit pillars in a church or temple. Evening light like this does even more to reveal the uniformity of Boardman's planting; each individual an equal distance from the next, branching out into leaf at the same level as its neighbour. I wonder what John Stewart Collis would have made of such evenly spaced trees, having enjoyed returning order to the overgrown ash wood in Dorset. As someone humbled both by nature and the agricultural manipulation of it, I think he would have found much to appreciate in the achievements of this farm.

Following breakfast the next morning, my coffee is kindly refilled in that enviable American custom. As I look out once again, downstream along the Columbia,

I notice a dot appear in the distance. After some time, the dot eventually reveals itself as a large industrial boat, chugging its way slowly upriver. It is of the kind built, like much of the curiously shaped machinery of agricultural America, for specific industrial purpose. Its flat metal framework displays a significant girth, and I ponder its intended function. (It isn't until the next day that, while sitting in the same spot, I witness a similar boat coming back the other way, laden with cut lumber.) The image brings to mind a trip I once made from Boston to San Francisco, so much of it on roads following or traversing the paths of great rivers. I remember crossing the vast Mississippi in Iowa and swimming the Boise in Idaho. In Wyoming the Shoshone led me into the valleys of Yellowstone, and the Niagara down along the Canadian border. During the frontier years of the early 19th century, rivers such as the Snake and indeed the Columbia were followed in pursuit of the West; they acted as navigational guidelines, creating a path through the labyrinth of country. Francis Parkman Jr., documenting his Oregon Trail experiences in the 1840s, noted these journeys as often commencing in the 'rapid current of the Missouri'[6], the other end of the Great Plains. Along my particular journey from east to west, I found the best way to make sense of the diverse and continually changing country was to note down the plant communities found from one state to the next. It was in doing this, especially while following the rivers, that I became familiar with America's poplars; those loosely collected under the moniker of 'cottonwood'.

Cottonwoods are so named for the white fluff of their seed casing, which, when ripe and loose from the tree, float easily on the air, often carrying great distances. The black cottonwood (*P. trichocarpa*) and 'necklace' cottonwood (*P. deltoides*) poplars together span the bulk of the entire United States, populating its river edges and saturated mud banks. It is a romantic name for a tree, frequenting the pages of many Great American novels. A fallen cottonwood ends the tragedy of Steinbeck's *The Grapes of Wrath*, and thickets of the trees conceal Huckleberry Finn along the Mississippi. Throughout my journey to Boardman, along the I-90 through Washington state, black cottonwoods appeared regularly, often the only tree to break through an otherwise exclusively coniferous landscape. I noticed them shimmering along rivers below undulating forested mountains, and thrusting from the silt of high, secluded lakes. If I had travelled on to Utah, south-east of Oregon, I might also have encountered the largest poplar in the world, so large in fact that it is a forest in itself. Forming a 100-acre (40.5ha) grove of over 40,000 trees, the suckering tree is believed to sprout from a single root system. 'Pando', as it has been named, is considered among the oldest, and largest living organisms on Earth, a feat far beyond the reaches of California's giant redwoods. Despite its homogenous root system Pando's outline shares much with Boardman Tree Farm. Its grove is dominated by a single species of poplar

(although it should be noted that Pando is an aspen, *P. tremuloides*), giving the appearance of a monoculture forest, similarly residing by an isolated road through the West. As such, the poplars at Boardman are no strangers in the territory, a distinctly American tree of significant cultural heritage.

In our last conversation about the farm, Austin likened its set-up to that of a giant hydroponics system. 'That's perhaps a better way to understand why the setting is so ideal here.' Hydroponic propagation, for the uninitiated, is a method of growing plants using a solution in place of a soil culture. Plants (often quick turnaround crops, such as salad leaves and annual herbs) are floated in rafts on the water with their roots dangling directly into it. The water is fed with a careful balance of nutrients offering the young plants all they require for swift, strong growth. In effect, hydroponic systems produce a synthesized environment conducive to efficient, large-scale plant propagation. Turning to Boardman, one might assume an absence of nutritional qualities in the arid soil of the Columbia Basin. Its inertness, however, is what makes the soil desirable as a growing medium. Like the water of a hydroponic system, the soil is a stable entity into which can be fed exacted nutrient measures. There is little to contend with by way of existing minerals which might otherwise complicate the feeding process. Just as the absence of regular rainwater led to controlled watering through irrigation, Greenwood Resources are also able to supervise a feeding regime under which their poplars can thrive. With these two primary pressures alleviated and no longer a factor, the climate in central Oregon actually becomes ideal. 'The days are reliably hot here and the nights suitably cold,' Austin concludes, 'the seasons are markedly defined'; conditions not too shabby for a temperate, deciduous forest.

I get the impression, by the end of my stay in Boardman, that its residents are used to avid interest in the farm. When asked about my visit it seems quite acceptable to enquirers that its purpose should reside with the trees – they're only surprised at the distance I've journeyed to see them. On one occasion a young waiter at the River Lodge and Grill responds with an anecdote of the time he ran a 15K charity run through the farm. He tells me 1,200 people took part, running in groups through the long avenues. That must have been a surreal experience, I say, trying to picture the wood so populated. The waiter describes how the forest felt tame with so any people inside, compared with experiences he'd had there alone. And this leads to a second anecdote – second-hand from a friend – of a cougar spotted prowling between the trees. If there are any cougars in there, I think to myself, how similar their perspective to Grand Coulee's confused salmon. Both iconic creatures of the wild, both trapped, in a sense, in the cross-lands between man and Nature. But as I myself can testify, and in fairness to the poor cougar, Boardman Tree Farm can appear deceivingly natural.

Beech

Common beech (*Fagus sylvatica*) is one of the temperate world's most admired and valued trees. As single specimens they can be enormous, both in height and crown, yet when grown as a low hedge they remain equally attractive, responding well to being clipped into tight shapes. Beech bark retains its smooth finish even in old age; it rarely fissures, often giving a mature trunk the appearance of an elephant's leg, splaying out at the bottom into roots frequently visible above ground. There are roughly a dozen species of *Fagus*, two of which are found in North America. The American beech (*F. grandifolia*) is a shade-tolerant, middling to large tree inhabiting forests on the east coast as far north as Nova Scotia; while *F. mexicana* grows in central and northern Mexico.

As beech trees tend to form a closed canopy, their woods are like no other kind: dark and atmospheric. However, around the middle of spring, when bluebells are in vivid flower throughout the beech woods of southern Britain, few forests are as overwhelmingly beautiful or as colourful as these. Transformation also takes place in the leaves of beech: in spring they begin in lime; by summer a dark green; in autumn they become golden yellow and in winter fall to the ground a rich, deep brown.

Beech (*Fagus sylvatica*)

Over ten years ago, a clear and simple vision united the founders of The Bushcraft Company: children should experience the fundamental wonders of the woods and have the opportunity to be immersed in this both elemental and educational environment. In this spirit, the outdoor education company grew rapidly from a small team offering a handful of summer-school camps to a nationally renowned operation working in multiple woodland sites across the UK. 'The first two summers were very special,' recalls Alice Hicks, former operations director and partner in the enterprise. 'My university degree results came out while I was camped in a pinewood forest in northern Scotland – all my friends were out partying and I was living in the middle of the woods, sat beside a fire with a group of school kids.' In the beginning, Alice and her colleagues approached schools fortunate enough to possess their own stretches of woodland, spending the summer months moving between camps and adapting activities to suit the various sites. There were wide games, challenges, wild swimming and tree climbing, together with exercises in fire lighting, foraging and wood carving skills. 'We named it The Bushcraft Company but in practice we were more about contact with nature as a whole; not so much survival techniques as a general grounding in the wooded outdoors.' The summer camps expanded quickly during the company's formative years, soon offering services to a number of schools across Britain. Being restricted, however, to those with pre-existing woodland access resulted in a narrowed demographic reach, and soon the company was looking for ways to extend its experiences to children from all backgrounds, particularly those at inner-city schools for whom woodland was not accessible. 'That's what led us to approach Cornbury Park,' Alice explains, a remnant of ancient woodland in Oxfordshire only a short distance from many major towns and cities. The idea was to set up a more permanent camp (to have the schools come to the wood, rather than the other way around) by basing their activities in the undisturbed beech woods of the Wychwood Forest.

Climb a Tree

'There is no question as to the loveliness of Cornbury Park. Its turf; its ancient earthworks; its glorious trees,' wrote Wychwood local, John Kibble, in his 1928 account of the forest and its villages. 'The gently rising ground, gradually falling away to where the water glints through the trees.'[1] Much like the New Forest in Hampshire, Oxfordshire's Wychwood was once a vast expanse of land designated as a Royal Hunting Ground. It was divided by the 1300s into three grand estates, including Cornbury Park near the town of Charlbury, which today encloses 5,000 acres (2,023ha) of farmland, gardens and sprawling broadleaf woodland. While little remains of the Wychwood's original wooded footprint, centuries of private ownership have rendered Cornbury's central tract relatively well preserved. 'I'd never before been in a woodland as beautiful and tangibly ancient as at Cornbury,' Alice recounts. 'There were huge trees to climb, lakes to swim in and big open beech woods to run beneath – it was idyllic as a setting and offered privacy for the camps.' Responding positively to The Bushcraft Company's enquiry, Cornbury's owners (the Rotherwick family) agreed terms by which their summer camps could take place in specified woods within the park, and before long 7–16 year olds from a host of urban settings were free to become feral among a captivating floral landscape. Alice describes a pattern of behaviour that would reliably unfold across the five-day courses at Cornbury; of kids arriving agitated and often tetchy, yet by midweek becoming relaxed, calmer and at ease in the woods. 'People talk about "green therapy" and the restorative qualities of forests, but in the Wychwood there was a definite sense of reinvigoration.'

In the early 1900s, 60 or so years before it was purchased by the Rotherwicks, Cornbury Park and adjoining house belonged to Vernon Watney, one-time High Sheriff of Oxford and dynastic heir to a beer brewing family. As proprietor of the park, Watney contributed to the records a beautiful leather-bound chronicle bringing together archived documents, photographs, maps and information pertaining to the estate's history. It is a giant, weighty book with a wandering narrative; a sort of glorified scrap book, privately published in 1910 as an enduring reference for the estate. Only 100 copies were produced of Watney's *Cornbury and the Forest of Wychwood*, though thankfully one of these resides at the British Library. Among its many appendices there is a substantial survey of Cornbury's trees, conducted in autumn 1907. It lists 5,406 in total, the vast majority made up of hawthorn, which is known to be a common feature in the understorey of the Wychwood Forest. Of the bigger trees noted however, the key players appear to be oak (both sessile and English), common lime, English elm and beech – the latter numbering 285 individuals. A few pages over reveals a photograph of a particularly stately beech – an inscription reads, 'King beech, blown down in September 1903'. Watney had the tree measured at a colossal 120ft tall with a girth of 21ft 4in. 'Rings about 230', he notes. When The Bushcraft Company first began operating summer camps at Cornbury one of the activities, as Alice remembers fondly, involved climbing the park's enormous 'Queen Beech', as it had been called; successor, perhaps, to Watney's king.

Naturally, matters of health and safety took precedence for The Bushcraft Company. There was a balance to strike with the wildness of the wood; children needed to connect with nature yet remain safeguarded within it. Guidelines therefore ensured that campsite tents and awnings were never erected below the park's mature woods, and regular assessments were made of the campsite trees and wider surrounding forest. Of all the people involved with risk-assessing a woodland for child activities however, few will have had the grounding that Alice herself experienced growing up. Her father, Ivan Hicks – garden designer, arborist and former presenter on *Gardener's World* – had been head gardener at numerous gardens in the south of England, including West Dean in West Sussex with its considerable and renowned arboretum. 'I'd spent my whole childhood in the woods,' says Alice. 'In the summer we'd decamp from the house to the garden, staying outdoors, playing in tree houses, following my dad around at work.' She remembers a giant redwood that she and her sister would regularly climb, holding tournaments with other children on the estate to see who could scale it the quickest. 'We live in a far more fearful age now. With parents generally involving their children less in outdoor activities, diminishing parental supervision has led to an overall lack of interaction with woods.'

It was the American journalist Richard Louv who, with the publication of his 2005 bestseller, *Last Child in the Woods*, coined the now familiar term, 'nature-deficit disorder'. The book argued that a childhood 'alienated' from nature is one that risks a host of psychological and physical deficiencies. Louv's findings turned heads across America but also in the UK, spurring the National Trust to commission the Natural Childhood Report, to assess the link between children's health and their contact with Nature, examining factors like technology advancement, increased screen time, rising obesity levels and parental responses to child safety. The organization subsequently launched a national campaign, heavily publicized in the press, listing '50 things to do before you're 11¾'. 'Make a mud pie!', it proposed, 'Dam a stream!', 'Go swimming in the sea!' Number 1, of course – placed at the tip of the spear – was simply to 'Climb a tree'. Where at one time Trust properties up and down the country might have asked visitors to keep respectfully to the paths, a revised outlook now encouraged arboreal explorations with refreshed, vertical horizons. A new webpage was added to the National Trust website: 'How to climb trees', citing a list of top tips and an inspirational video for good measure.

Why is it, then, that we view climbing a tree as the pinnacle of interaction with Nature? Why should it rank above daisy chains and rock-pooling in the mind of the disaffected urban public? There's little question over connotations with the feral and the wild – I think again of John Muir swaying high upon his windswept Douglas-fir – but there's more to it than that. Tree climbing has to some extent always been an act of rebellion. The National Trust's campaign was so effective, I believe, because

it posed such a dramatic reversal of their former buttoned-up approach to the management of natural, heritage landscapes. It was a dare, almost; an appeal to a childlike rebelliousness: 'Go on; we won't mind, climb our trees'. But central to the exercise is an interaction with an organic, unregimented entity; it is a physical dialogue with nature that, ultimately, offers an attractive alternative to the restraints of modern living.

Balkan Beech Woods

Two things come to mind following a morning walk at Croatia's Plitvice Lakes. The first is that its beech woods, sprawling high above the water across the surrounding hills, are, in colour, shape and form so much like those back home in southern Britain. And the second, having witnessed at the lakes themselves a landscape as close to 'heavenly' as any I've encountered before, how could such a staggeringly beautiful setting have once played host to the brutality of war? It's hard to imagine violence breaking through the serenity of Plitvice's paradisiacal waterfalls and turquoise pools, but in March 1991 the area saw a fatal confrontation, the first in Croatia's war of independence from Yugoslavia. The four-year conflict prompted UNESCO (the United Nations Educational, Scientific and Cultural Organization) to place the Plitvice Lakes, already a national park and designated site of outstanding natural beauty, on its list of World Heritage Sites in Danger. Preserved and protected, the area thankfully remains one of Europe's most spectacular natural treasures; 16 iridescent lakes divided by waterfalls and chalky travertine deposits, watched over by dark forests of beech and fir.

Where their leaves are concerned, beech trees are afforded two great 'moments in the sun', so to speak. Their autumn foliage is among the best of the season's arboreal colour, blazing yellow-orange against smooth, unfissured bark. But beech leaves are to me, as they are to so many, the true leaves of spring. They are among the first to absorb the new season's warmth; the first to break bud and radiate a fresh green through the waking forest. All new broadleaf tree leaves open in soft, spongy contrast with their papery end, but the thin, almost translucent leaf tissue of brand new beech is of a considerable superiority, its chlorophyll glowing under the skin like torchlight in a tent. Thomas Hardy once described their young, unfolding form as possessing the 'softness of butterflies' wings'. Like butterflies they are a display of momentary splendour, exhibited in that brief window before the canopy catches up and the forest darkens for summer. This is why I have come to the woods above Plitvice National Park; to witness this spectacle in one of southern Europe's most impressive and exotic landscapes.

The forests surrounding the Plitvice Lakes make up a staggering 79 per cent of the park's 183-mile (295km) land mass and extend much further beyond its boundary. This means that what isn't water or walkway is almost exclusively forest, a virgin landscape barely altered by man. One passes along the narrow footpaths surrounding the lakes as if suspended in air, floating lightly through a landscape of water and wood. There are plenty of tree species to be found in the park, including sycamore, spruce, wych elm and maple, and at the water's edge are a great many hop hornbeams (*Ostrya carpinifolia*), slinging out low branches over the motionless lakes. On account of their similarly shaped twigs, buds and leaves, these trees are often confused with beech. Beech however – much taller and with smoother foliage – is the overriding tree at Plitvice, most of which are grouped together in great, dense swathes. Higher up they intermix with silver fir, forming a separate, wilder woodland ecology. These forests are of an entirely different tone, home to southern Europe's more elusive mammals, such as wolves, lynx and the European brown bear.

As mentioned at the outset, the woods predominantly of beech conform in many ways to those of familiar downland England. The eye-catching similarity is the presence of calcareous rock. Travertine, much like downland chalk, litters the woodland floor; here and there lumps of brilliant white protrude through the previous year's fallen leaves. Travertine is the name given to the sedimentary rock deposited by mineral springs, found often at the summit of steaming geysers and underground limestone caves. As with chalk it is an alkaline substrate, impressing upon the lakes their crystal clear, turquoise water. The calcareous deposit is visible at the base of the dams, themselves the result of hardened barriers of built up travertine. In Italy, as in the many places it naturally occurs, travertine was excavated for use as a construction material, a precursor to modern day concrete. The Colosseum in Rome is a lasting representative of the substrate's attractive application, contrasting strikingly with the surrounding stone pines. But to return to Plitvice, this permeating alkalinity in the region creates an environment in which beech can thrive. The common beech, *Fagus sylvatica*, will tolerate all manners of soil, provided it is free draining. However, gazing between the silver-grey trunks rising from an

uneven, leaf-mould floor, and at the lumps of white stone scattered throughout Croatia's elevated forest, I find myself thinking of the Sussex Downs, and their homogenous, tranquil beech woods. The journey to Plitvice, beginning at the coastal town of Split, offered another association with plants from home. Heading first to the Krka river further south, I experienced that horticultural delight in encountering familiar garden plants growing wild in their natural habitat. In the arid hilltops of Brištane to the east of Krka National Park, airy plumes of smoke bush (*Cotinus coggygria*) shimmered in the sunshine like pink-tinted spiders' webs. *Cotinus* was one of the first shrubs I got to know while studying at the National Botanic Garden of Wales and has remained a favourite ever since purely for this reason. Its papery, disc-shaped leaves used to catch my attention walking through the gardens each morning, spreading beside a stone wall in cloddy Welsh soil. Along Brištane's roadside verges and rooted into the hillsides were countless *Cotinus* shrubs, flowering freely in growing conditions much more favourable than those of damp rural Britain. Similarly, closer to the river I came across *Nigella damascena*; a wild flower I've so often sown from a seed packet but there found feral, colonizing loose stone and grit with its jagged blue petals.

Similar to Plitvice, although unequal in bewildering scale, Krka's river course flowed over waterfalls and into pristine lakes, underpinned by a travertine basin. Grey wagtails and white campion flowers resided along its banks, and fruiting fig trees abounded beside ash and black poplar. Unlike the beeches in Plitvice's deep and earthy forests, ash, a tree of parallel if not firmer familiarity, appeared almost tropical at Krka. At every turn I met the dangling foliage of either its common or flowering species, draped like willow across winding paths or dominating the little islands that slowed and parted the river. It was satisfying to observe ash trees as the key players in a landscape, given their continued threat from fatal dieback disease across the European continent. The tree, reflected in tranquil clear waters teeming with placid fish, roused an uncomfortable vision of the ash having died and gone to heaven in this idyll of natural beauty.

Krka in many ways set the tone for Plitvice; an introduction to the wild terrain of the Balkan Peninsula. It failed, however, to prepare me for the theatrics of its temperamental weather. On the penultimate day of the trip, having crossed one of the lakes to revisit a stretch of mature beech wood, there came the rumble of clouds and the advance of heavy raindrops. The clatter was quickly amplified as rain turned to hail and I made a swift dash for cover under the wide, arching canopy. This unique event caused light to drain from the wood almost as if dimmed with a switch. Beech trees are often referenced as forming especially dark woods, and here they were no exception. I stood for a while adjusting to the gloom, keeping dry at the base of the thundering forest. An uneasiness settled as the trees became taller, their upper branches disappearing into an engulfing darkness. Every now and then muted flashes of faraway lightning lit up the damp tree trunks, reflected on the sheen of their wetted skin. A little further up the hill there loomed an exit from the woods in the form of a light at the end of a short tunnel. The passage led to an elevated position above the lake from which I could view the rainfall and gauge its duration. While remaining under the relative cover at the fringes of the forest I waited for the downpour to pass. Had the thunderstorm not occurred however, I might not have been drawn towards this particular spot and I would have missed a crucial find. In the hillside glade, growing out from layers of leaf mould, I came across two hidden gems. Both were flowers of striking colour and both indigenous to beech woods. Aptly named, given the circumstances, they were bastard balm (*Melittis melissophyllum*), a pink and white lamium, and yellow melancholy thistle (*Cirsium erisithales*), a member of the sunflower family. This I learned from later identification, as at the time they were new to me. The thistle in particular, a lover of calcareous soil, was radiant beyond anything typical of a beech wood. Here in a forest so otherwise markedly familiar, grew something wonderfully alien, brought to my attention by Mediterranean thunder.

Cherry

Three *Prunus* species are native to Britain and Ireland; the low, shrubby blackthorn (*P. spinosa*), which produces sloes, the wild cherry and bird cherry. Perhaps rather confusingly the scientific name of wild cherry or gean, *P. avium*, implies a relationship with birds, while the bird cherry is named *P. padus* (Greek for wild cherry). The Latin name of the former relates to dispersal, as the fruit of wild cherry is avidly eaten by birds and its seed is subsequently spread in their droppings – hence the name *avium*. Distinguishing between the trees themselves is slightly easier. Wild cherry trunks are ringed horizontally with cream-coloured lines, and their blooms are prolific and distributed in clusters, similar to apple blossom. In contrast bird cherry bark is a dull grey-brown and its flowers are tightly grouped, finger-like racemes. Worldwide, *Prunus* is a group to which many familiar fruit trees belong, such as almonds, peaches and apricots – those bearing fruit with stones at the centre. Varieties of plum, such as greengage and damson, originate from subspecies of the widespread *P. domestica*.

Bird cherry (*Prunus padus*)

Hanami in Copenhagen

For a second time the subject of this book turns to an arboricultural practice with its origins in Japan. A thread seems to be woven through Japanese custom and culture that associates trees, in one form or another, with all stages of life. *Hanami* celebrates rebirth through the transience of spring blossom; *shinrin-yoku* is a kind of arboreal meditation intended to alleviate the pressures of modern living; and in several Japanese burial sites trees have replaced gravestones, assisting the departed in feeding directly back to the natural world. Buddhism – still by far the most commonly practiced religion in Japan – accounts for a proportion of this ethos; a legacy born at the Bodhi Tree under which Buddha sat in contemplation. But in modern-day Japan sylvan appreciation straddles both the spiritual and secular spheres, seen perhaps most definitively in the arrival of cherry blossom each year. Between March and early May a great wave of flowers sweeps up through the nation's islands, from the south to the north, unfolding in the street trees of Nagasaki and Tokyo, and in the wild cherries of Honshu's hill forests. The event is monumental, not just for the sheer volume of trees clothed in brilliant white and pink, but for the cultural interaction that takes place beneath their branches.

A single word denotes the splendour of cherry blossom in Japanese; *sakura*, and contemplative admiration of *sakura* is thought to have been a fixture in the country since the 8th century. It is a ritual in which people gather under cherry canopies to marvel at those particular and wonderful inflorescences produced by the ambassador of the *Prunus* dynasty. With their springtime arrival and short-lived display, the flowers are considered to be a physical representation of life's beautiful yet ephemeral nature, and today they are rung in with parties all across the country; an observance known as *Hanami*, celebrating the season of renewal. Street food vendors set up tables under the trees; workers finish early to picnic in parks with friends and family; and by evening the blossom is illuminated with bright filtered lights so that celebrations can continue late into the night. Rising *sakura* tourism has necessitated a national blossom forecast that predicts the date of peak display for each region of Japan. The window is relatively narrow – cherry blossom at its best tends only to last a week or so – and as festivities can often take over whole streets and town centres, organizers rely on this information to ensure their events coincide with the flowers.

But while fluctuations in weather dictate that its timing will vary year to year, *sakura* itself is one of nature's reliable annual fixtures, the natural equivalent of fireworks on New Year's Eve. Few events in the calendar connect civilization with the Earth's movements in such a profound way, and fewer still are celebrated with this level of expectancy and delight. Many people, like myself, await the return of migrating birds each summer, like the swifts that come bounding back into London following winters in Africa; similarly, snowdrops are cheered by all gardeners as the first of the bulbs flowering in early spring; but *sakura* blossom displays a sense of both transience and delicacy from within a city context, enacting a metamorphic transition in an environment that would otherwise change very little. It offers an 'elemental presence', as Henry Beston might have put it, that is rarely experienced these days. 'The world today is sick to its thin blood for lack of elemental things,'[1] the American writer declared almost a century ago, a sentiment as true now as it was in his time. And in a visually driven society, where images are shared as easily as they are taken, the event is well documented and observed by an unparalleled audience. It has spread, in fact, across many continents, including North America and Europe, where cultivated varieties of Japanese cherry (*P. serrulata*) have been planted for their *sakura* spring effect. Cities like Washington D.C., Vancouver and Stockholm host their own *Hanami* events, which typically involve festivities in celebration of Japanese culture.

For over a decade *sakura* has arrived on Copenhagen's eastern waterfront in the cherries of Langelinie Park. The 200 trees planted in the grounds were gifted by the Danish Honorary Consul in Hiroshima in 2006, and the following spring saw the first of many *Hanami* festivals conducted beneath their blooms. Denmark and Japan share a good diplomatic relationship, stretching back to the Treaty of Friendship, Commerce and Navigation established between the two countries in 1867. As 2017 was the 150th anniversary of the treaty, the Japanese embassy in Copenhagen ensured that Langelinie's annual *Hanami* festival was more exuberant than ever.

Arriving in late April for the weekend event, I join hundreds of the city's revellers under a shower of pink-white petals. Percussive traditional music blares from speaker stacks at the top of the park, where a stage has been erected to host a variety of performances. There are demonstrations of judo, ninjutsu and karate, a reading of haikus by a local poetry group, and a few songs belted out by a Japanese pop singer with a Casio keyboard. Authentic cuisine, tea ceremonies and origami classes feature between the trees too, dispensed from canvas awnings and long, decorated tables. A samurai warrior in customary dress wanders through the crowd, stopping every now and then to pose for Instagram photos. The festival as a whole is a curious sight to witness. So much liveliness is squeezed into a surprisingly small space, kept rather neatly to the interior of the trees; families spread blankets down on the ground and enjoy the spirited atmosphere, rained upon by intermittent petal fall like a lacklustre snowstorm.

Langelinie Park is a good spot for a spring festival. Perched on the edge of the city it faces the Øresund channel – a waterway separating Denmark from Sweden, flowing between the North and the Baltic seas. A light, fresh breeze therefore sweeps in and under the trees, and sunlight bounces freely from the water embellishing their figures with a reflected light. By the early afternoon however, with further crowds arriving for the festivities, I decide to head across town for the quiet of Bispebjerg Cemetery. Bispebjerg sprawls out from the north end of Copenhagen and, like many expansive graveyards, possesses a large number of tree species, including a notable avenue of ornamental cherries. As the city's youngest burial ground, its municipal foundations mostly follow the shape of a systematic grid, a structure that lends naturally to the implementation of trees in rows. As such, Bispebjerg doubles as a kind of arboretum of avenues whereby each pathway is banded by its own line of trees. Walking up from the main gate I wander idly in a peaceful zigzag, swapping long rows of hornbeam and oak for cedar, beech and pine. Smaller paths in between have been edged with birch and katsura or darkened by the heavy matting of *Cupressus* foliage. The layout here bears a resemblance to the Boboli Gardens in Florence – labyrinthine corridors of planting designed to tower impressively above the eye line. An increasing arboreal inquisitiveness propels me forwards from one section to another and led in this way towards the top left corner I at last spot a bright tunnel of cherries, radiant in full flower. They are an avenue known well to Copenhagen locals, but, with the hordes occupied this weekend at the other end of the city, for a moment I take in the serenity of the *sakura* by myself.

A morbid and fearful tale looms over Wayland Wood in Norfolk. Abandoned deep in an eerie woodland by a murderous aunt and uncle, two orphaned children wander hopelessly lost until, eventually, they starve to death. *Babes in the Wood*, as the legend is memorably titled, concludes yet more sombrely with the bodies of the children lying unburied and forgotten, 'till Robin Redbreast painfully did cover them with leaves'. The story has existed since the 16th century and has, like all traditional folklore, accrued many different versions across the years. But such is the babes' legacy that, however told, their tale remains inextricably bound with Wayland; a 96-acre (39ha) enclave of ancient woodland in the heart of East Anglia. Even its name – an adulteration of the original Old Norse designation, Wanelund – over time became 'Wailing', as, according to the legend, the children's ghostly cries still haunt the trees at night. However, this is comparatively recent history for a wood with records dating back to Anglo-Saxon times, and it is a shame, in a way, that such sinister associations override what is actually a unique and attractive woodland. For a start, woods are far from a common feature in the topography of East Anglia, restricted to isolated pockets that sit like islands within a flat, arable landscape. For this alone Wayland Wood is a curious sight, even more so as a remnant of ancient woodland, but the trees themselves, along with the resident ground flora, ought really to hold the limelight.

The predominant tree species are ash, maple, oak and hazel, the latter two having been managed for centuries as coppice with standards for the production of timber. However, atypically for this eastern region of Britain, bird cherry (*Prunus padus*) is also present and prolific, forming a key component within the structure of the wood. The tree mingles among the hazel where it is similarly coppiced and can be found throughout as single specimens rising between the shrub and canopy layers. The cherries are one of Wayland's many intriguing constituents, thought to thrive on account of the wood's favourable sandy-loam soil structure. But there are other oddities too: the yellow star-of-Bethlehem flower (*Gagea lutea*) for example, which in all of Norfolk has elected to grow only here, and the mysterious absence of dog's mercury (*Mercurialis perennis*) that, as an indicator of ancient woodland, ought to flourish in the leaf-litter humus. And then there are the

Wayland Wood

dead things; for even without the *Babes in the Wood* connotations death is a prominent characteristic of Wayland. In 2014, fresh from the continent, ash dieback disease found its way to the wood's ash population, creeping in at the leaves and decimating top growth to defoliated, blackened rods. Now well and truly established, the destruction is nearing its zenith and once elegant, well-managed trees are beginning to succumb at the trunk and fall. This isn't so much from ash dieback directly, but weakness is preyed upon quickly in the natural world and rot will finish what disease has begun; they serve the same function. Peel away the bark and you'll find rhizomorphic bootlaces of honey fungus, a black net cast up over the bole with which to pull it to the ground for decomposition.

Steve Collin of the Norfolk Wildlife Trust describes Wayland Wood as a 'fungal playground', a rich and damp habitat ideal for the spread of mycological organisms. As reserves manager for the Trust, Steve has looked after the wood for over 18 years. He was in fact the first to flag the signs of dieback in its ash trees and worked with the Forestry Commission to authenticate the disease. I arranged to meet Steve in Wayland for a visit in late February, to be introduced to its cherries and get a picture of how their unusual occurrence plays into the overall structure of the woodland. Early on a midweek afternoon we meet by the entrance and, in the wintry air of an approaching cold snap, Steve helps me decipher the wood's intriguing contents. But there is something else; in spite of my playing down the significance of Wayland's famed fable, something in me couldn't resist an urge to explore the wood at night. There are two reasons for this. First, few of my forest experiences have yet allowed for a nocturnal visit, and a wood, as is well documented, is a different entity under the cloak of darkness. Second, and perhaps led by a kind of underlying nostalgia, I craved something of that unnerving and childish sensation of walking alone through a forest in the dead of night. Therefore, I have brought along a tent and a few bare essentials for an overnight stay, making arrangements with the Wildlife Trust to let me occupy a spot next to the trees. But as I begin my walk with Steve through the central riding of the wood, the leafless, frigid twigs and soggy earth reminding fervently of the season, I can't help thinking I've made a poor choice – not only in timing my nocturnal excursion for the middle of winter, but in locating it in Britain's most notoriously haunted wood. 'How endlessly beautiful is woodland in winter,' wrote Gertrude Jekyll – but she probably didn't contemplate spending a night camped out in it.

Bird cherry's most distinctive quality is revealed when you crush one of its young stems; an astonishing marzipan aroma rushes out, heavy and sweet. Cherries belong to the *Prunus* genus, which also includes the wild, edible cherry (*P. avium*) as well as the almond (*P. dulcis*), and almond is very much the aroma of cut bird cherry, as Steve points out beside an area of Wayland's coppice. The scent makes you want to eat it, though this would be unadvisable, Steve says, on account of a low level of hydrogen cyanide residing in the sapwood. The poison is what makes bird cherry an excellent tree for coppice, as, unlike the new shoots of hazel, browsing mammals – deer and rabbits – are deterred somewhat by the foul taste. Other than this, cherry coppice is very similar to hazel. Both exhibit that sheeny bright bark, polka-dotted with tiny lenticels, and their buds are similarly positioned in an alternate pattern, pointed outwards along the stems. Cherry stools also take an equal level of butchery when cut back, and the resulting regrowth is as compact and durable as hazel.

As a wood, Wayland is surprisingly compartmentalized and navigable, divided into sections made accessible via wide, cleared ridings. At one stage the whole wood was managed as coppice and records show that, during a few ten-year periods, Wayland was felled in its entirety. 'This was 1600BC,' Steve says, '"before chainsaws" that is, which, when you think about it, would have been a staggering achievement.' No doubt there was a high demand for the timber extracted during this time, relied upon for countless uses in the surrounding farmed landscape. These days about a quarter of the wood is coppiced, which is still an exemplary proportion, and as a result Wayland reflects a tangible sense of its orderly, former industrious employment. Towering above the coppice are remarkably straight and broad standards of oak and ash. They are evenly spaced through years of careful management and allow a good level of light to fill the understorey. It is a sting to think that such well-looked-after examples of ash are already in decline from dieback disease, and that no good timber will come of their many decades of careful husbandry. Outside the coppiced areas substantial parts of Wayland are designated as minimum intervention zones – sections of the wood allowed to become 'messy' with natural decay and regeneration. Steve leads me into one particular division in which huge field maples and hornbeam limbs rise from colossal, ageing stools. These trees are the bones of a derelict coppice that, having fallen out of rotation almost a century ago, now lean and twist into broad and unlikely physiques. There are bird cherries in here too and, with the freedom afforded by the minimum intervention zone, they display their congenital habit of arching over and brushing the ground. Bird cherry naturally self-layers: once a branch makes firm contact with the soil it will put out new roots and begin as a new plant. It is linked to the parent tree in a sort of subterranean daisy chain. Steve and I rummage through the low branches of one individual and, sure enough, we find a fresh young rod tethered to the earth by a set of its own roots.

The rest of the afternoon is spent uncovering as many of Wayland's curious components as time allows for: small ponds enclosed in scrub; tangles of rampant honeysuckle and the bright red rubber of scarlet elf cup fungi, sprouting from a branch sodden with decay. And when the light begins to redden towards the west with a sinking, wintery sun, Steve and I return to the wood's entrance where he wishes me well for the night ahead. 'Some people say the wood is quite eerie,' he concludes. 'But then all woods are at night – they clunk, grind and bang – though obviously the *Babes in the Wood* history gives Wayland an additionally creepy edge.' We say goodbye and I promptly busy myself with setting out the tent; this way I'm committed for the night, however creepy the edge turns out to be. To protect Wayland's rare and ancient heritage, the Wildlife Trust do not allow anyone to camp in the wood itself but Steve has kindly organized for me to pitch beside the outer trees on private ground, so that I can slip in and out from the perimeter with ease.

Tent set up, torch and notebook stuffed in pocket, I head into the wood and trace a path towards the middle. It's 5.30pm and I make it back just before the light has all but drained from the trees. Positioned at the base of a large field maple I stand still for a moment, hold my breath and listen. An extraordinary amount of bird activity seems to be reverberating across the wood. It's all chatter and clamour as disputes are resolved and perches reclaimed; thrushes, blackbirds, a particularly noisy robin too – perhaps the fabled undertaker come to bury me with leaves. And then something I've not experienced before, or at least not in so perceptible a way; a sudden quietening down and abrupt fall into silence – just as the clock hits 6pm. It is as though the birds have, as one, adhered to some formal code and turned in for the night. Only a hapless cock pheasant, unschooled in British woodland

survival, gives away his position with a staccato crow like a failing car ignition. The light is now a waning gloom and dusk is coming to a close; all trees lose their outline, save for the gilded silver of young hazel rods in the coppice. Ash bark also remains visible a little while longer with its duller, warmer hue. I sense my eyes adjusting to hold these images but soon enough the wood is cloaked in darkness.

A wood at night is a notorious thing. In literature, both in fiction and non-fiction, it is the domain of the unexplained and a stage set for fearsome events. Of the latter a particular favourite is a passage in *Through the Woods* by H. E. Bates describing experiences of woodland after dark. He writes of being caught out while gathering violets and primroses late in the evening, 'when the sudden realization of twilight coming down has sent a sudden damnable running of cold up my spine, and I have half run out of the place.'[2] What I like most about the passage is Bates's attempt to unpick the particular characteristics that account for a wood's menacing night-time persona. 'It is not simply darkness,' he asserts with inquisitive deduction; 'we grow used to darkness.' Perhaps it is the claustrophobia brought on by a sky-less roof, he wonders, or the presence of poachers, or simply 'the seclusion of the place'. Ultimately Bates concludes that it must be something about the trees themselves; 'they impinge on us, hypnotize us,' he writes, and perhaps more profoundly; 'night changes them, or it changes us.' This, surely, is the crux of the matter; that our imagination is the primary culprit.

Around 8pm I return to the tent for some food but wander back for 10pm and find that new life has awoken in the wood and the mood feels different again. Up in the trees all remains silent but on the ground there are now rustles and voices. Mammals are up and roaming about, uttering yelps and distant barks. The wood also seems blacker than it was before, despite the glow of a crescent moon that has emerged through dissolving cloud. Rather than stay put, this time I elect to keep moving, following suit with the animals. A meander down the ridings takes me onto a path unrecognizable from the afternoon that borders a recently cropped section of coppice. Only children, I think to myself – and very small children at that – could get lost in a wood cut so low as this; the tips of the stems reach just above my head and in the light I could probably see over them. But then a figure moves out in front of me in a fluid stride and my heart is immediately in my mouth. The compact silhouette seems to have appeared from nowhere but, to my relief, I see it is a muntjac deer and, apparently unaware of my presence, it continues up along the path ahead. A small silver moth floats into the riding and passes beside me too – it follows behind the deer in a strange, silent procession. I smile at the thought of a faunal meeting, arranged somewhere in a corner of the wood, all creatures obeying the summons.

A conversation of tawny owls is the final instalment of night-time activity, sometime just before midnight. Spread wide across the wood they respond to each other with wonderful sporadic hoots, solemn yet assertive. By now the cold is gnawing at my bones and all I can think about is the padding of my sleeping bag, but it's a long walk back to the tent and, when I do reach it, the wood seems almost to close up behind me. It feels suddenly very good indeed to be out from under the trees and standing in the dim of the periphery; apparently imagination is at its unbridled worst when making a final exit from the wood itself. It's close to 2am when at last I begin to fall asleep inside the tent. An owl perched nearby has been calling at long intervals and, being so close, I've been absorbed listening to the full-bodied depth of his intermittent outbursts. Then somewhere on the drowsy threshold of sleep a sharp scream comes spinning up from the trees – a ghost? – a babe in the wood? Or just the territorial bark of a muntjac ... And there's that unnerving sensation I've waited for; there's the childish thrill – a rush of primal alarm that shudders in the shoulders and makes the heart jump. In the tent I slip slowly back into a weary daze, while outside the wild wood goes on being wild.

In the half-lit morning I unzip to a frosted world of sugared grass and whitened twigs. I knock a thin layer of snow from the tent before pulling it down, packing my rucksack and walking back along the ridings through the wood to where I'd left the car. Birds are up and singing at full volume again; among them this time the maniacal laugh of a green woodpecker, a feral echo of last night's activity. Below the canopy however, all is frozen and still – dormant, empty, soundless. Dead and weathered stems of burdock and cow parsley glint cold and hard below chandeliers of hazel catkins, and underfoot the snow lies clean and powdery, as yet unmarked by wandering animals. I find myself thinking of spring. In a patch of exposed ground, sheltered beneath an ash, a carpet of moschatel leaves prepares to bear their little globes of lime-green flowers. When this starting pistol has been fired there will be no turning back; all through the wood swollen buds will break into yellows, purples and white; one flower after another in dependable succession. There'll be anemone, primrose and celandine, then bugle and violet and hoards of carpeting bluebell. But at the tail end of spring, just when the great flourish seems over with, the cherries will have their moment too, whitewashing the wood for summer.

Photographer's Note

When Matt and I first pitched the idea for the book we were perhaps unaware of the enormity of the project. Each two-part chapter involved meticulous research before travelling to the locations and patiently waiting for the key shooting hours at sunrise and sunset. The people we have met have been wonderful and really have broadened our perspective with their thoughts on nature that they were kind enough to share.

My approach to this book has been to view this complex project as both artistic research and scientific enquiry. My aim was to experience and document the journey through the trees and share this via a collection of carefully captured photographs. No single image is intended as a 'hero shot', rather, a series that twists and turns with the text before meeting harmoniously with the reader.

I shot everything on 6×7 negative and made the enlargements in the darkroom. I chose this path as I feel the medium expresses our own intrinsic connection to nature. We naturally understand and connect to the image through the imperfect way film renders light – its essence is beautifully flawed. I was lucky enough to experience the landscape first in person and then again many weeks later in the darkroom. I discovered that this process of reflection allowed me to revisit not only the vistas, but the emotional nature of this journey.

Resources

Directory

Copenhagen Sakura Festival, Denmark
www.sakurafestival.dk

The Bushcraft Company, UK
www.thebushcraftcompany.com

The Botanist Gin, UK
www.thebotanist.com

Grain & Knot, UK
www.grainandknot.com

GreenWood Resources, Global
www.greenwoodresources.com

Oregon Natural Desert Association, U.S.
www.onda.org

Pacific Rainforest Adventure Tours,
** Vancouver Island, BC**
www.rainforestnaturehikes.com

Micologia Forestal & Aplicada (MF&A),
** Catalonia**
www.micofora.com

Norfolk Wildlife Trust
www.norfolkwildlifetrust.org.uk

Ruislip Woods Trust, UK
www.ruislipwoodstrust.org.uk

Trees for Life, UK
www.treesforlife.org.uk

Truffle Tour, Catalonia
www.tempsdetofona.com

Verderers of the New Forest, UK
www.verderers.org.uk

Wolf Watch UK
www.wolfwatch.uk

Useful information

Bedgebury National Pinetum (UK)
www.bedgeburypinetum.org.uk

English Woodlands Nursery (UK)
www.ewburrownursery.co.uk

Garden Museum (UK)
www.gardenmuseum.org.uk

The National Trust (UK)
www.nationaltrust.org.uk

The Wildlife Trusts (UK)
www.wildlifetrusts.org

Woodland Trust (UK)
www.woodlandtrust.org.uk

Footnotes

Ash, Mulberry and the Wild Wood
p.14
1 Richard Ingrams, *John Stewart Collis*,
  Chatto & Windus Ltd, 1986, p.102

Pine
p.23
1 Jim Crumley, *The Great Wood*, Birlinn
  Limited, 2011, p.59
2 Hugh Johnson, *Trees*, Octopus Publishing
  Group, 2010, p.88
p.27
3 Jim Crumley, *The Great Wood*, Birlinn
  Limited, 2011, p.xvii
p.36
4 John Stewart Collis, 'The Wood', *The Worm
  Forgives the Plough*, Vintage, 2008, p.234

Hornbeam

p.52

1 John Stewart Collis, 'The Wood', *The Worm Forgives the Plough*, Vintage, 2008, p.247

Douglas-fir

p.71

1 Roger Deakin, *Notes From Walnut Tree Farm*, Hamish Hamilton, 2008, p.151

p.86

2 John Muir, *The Wild Muir*, Yosemite Conservancy, 1994, p.109

Oak

p.106

1 Richard Jefferies, *Walks in the Wheat Fields*, Penguin Books, 2009, p.74

p.116

2 'Twenty-five years of change in a population of oak saplings in Wistman's Wood, Devon', Edward P. Mountford, English Nature, 2000

Juniper

p.124

1 Gavin Maxwell, *Ring of Bright Water*, Little Toller Books, 2009, p.24

p.130

2 J. A. Baker, *The Peregrine*, Harper & Row, 1967, p.143

p.142

3 Adam Nicolson, *Chasing Wolves in the American West*, Granta, 2014, p.131

Birch

p.157

1 Gertrude Jekyll, *Wood and Garden*, The Ayer Company, 1983, p.8

p.163

2 Richard Mabey, *The Ash and the Beech*, Vintage, 2013, p.xvi

Chestnut

p.177

1 Edward Thomas, *In Pursuit of Spring*, Little Toller Books, 2016, p.32

p.181

2 Thomas Hardy, *The Woodlanders*, Wordsworth Editions Limited, 1996, p.170

Poplar

p.196

1 Richard Mabey, *The Ash and the Beech*, Vintage, 2013, p.250

p.200

2 Kirk Johnson, 'Down the Mighty Columbia River, Where a Power Struggle Looms', *The New York Times*, 2017

p.201

3 Hugh Johnson, *Trees*, Mitchell Beazley, 2010, p.194

p.203

4 Richard Mabey, *Nature Cure*, Pimlico, 2006, p.207

p.204

5 Chris Starr, *Woodland Management*, The Crowood Press, 2005, p.7

p.208

6 Francis Parkman Jr., *The Oregon Trail*, Oxford World's Classics, 2008, p.8

Beech

p.217

1 John Kibble, 'Wychwood Forest and its Border Places', *The Oxford Chronicle*, 1928, pp.12–13

Cherry

p.236

1 Henry Beston, *The Outermost House*, Henry Holt and Company, 1949, p.10

p.247

2 H. E. Bates, *Through the Woods*, Little Toller Books, 2011, p.53

Index

Acknowledgements

While conceiving a project of this measure leant predominantly on the keen ideas and past experiences of a long-standing partnership, bringing it all together – into the physicality of a book – relied upon the guidance, expertise and unending patience of a great many people, to whom we are most grateful. Among those who have contributed directly to this book's content, supported its undertaking or simply pointed us in the right direction, we wish to thank: Lester Hillman, Doug Gilbert, Karen Mitchell, Dom Andrews, Richard Hutton, Corey Burant, Tony Haighway, Syd House, Ronda and Gary Murdock, Marcos Morcillo, Albert Boixader Faja, Stevan Martin, Carl Reavey, James Donaldson, Genessa Goodman-Campbell, Lace Thornberg, Chelsee Richardson, Sophie Sellu, Sue Westwood, Austin Himes, Alice Hicks, Steve Collin, Christopher Woodward, Wendy Leston and Daniel Luscomb. Thanks also to English Woodlands Nursery and to Bedgebury National Pinetum for the provision of tree leaves.

Thank you to all at Pavilion Books for supporting a truly enjoyable and challenging experience, in particular Krissy Mallett, Laura Russell, Claire Clewley, Katie Hewett and Katie Cowan.

Carl Reavey, who so kindly introduced us to the natural curiosities of the Inner Hebridean Isle of Islay, sadly passed away in January 2018. His loss is felt greatly by all at The Botanist and the wider island community, and his generosity, knowledge and enthusiasm will be remembered fondly.

Matt
Books tend to be approached in anticipation and acceptance of a certain demand on personal time and energy, however for that which is endured, inevitably, by others – family, friends and colleagues – one can only offer gratitude and a stack of IOUs. At the top of the pile are Clemmie Power Collins and Siân Collins, who not only put up with my absence, but helped steer the resulting text.

Roo
There is a long list of people to whom I am indebted for without their immense contribution and support this book would have been a project far too large to fathom. I hope all these incredible people know who they are but the special mentions go to my forever understanding partner Lauren Ingram, Fizz and Geoff, Tom Johnson and Chan Photographic Imaging.